The Rightful Place of Science:
Climate Pragmatism

The Rightful Place of Science:

Climate Pragmatism

Edited by

Jason Lloyd
Ted Nordhaus
Daniel Sarewitz
Alex Trembath

Foreword by

Calestous Juma

Consortium for Science, Policy & Outcomes
Tempe, AZ and Washington, DC

THE RIGHTFUL PLACE OF SCIENCE:
Climate Pragmatism

Printed in Charleston, South Carolina.

The Rightful Place of Science series explores the complex interactions among science, technology, politics, and the human condition.

For information on The Rightful Place of Science series,
write to: Consortium for Science, Policy & Outcomes
PO Box 875603, Tempe, AZ 85287-5603
Or visit: http://www.cspo.org

Other volumes in this series:

Allenby, B. R. 2016. *The Rightful Place of Science: Future Conflict & Emerging Technologies*. Tempe, AZ: Consortium for Science, Policy & Outcomes.

Cavalier, D. & Kennedy, E. B., eds. 2016. *The Rightful Place of Science: Citizen Science*. Tempe, AZ: Consortium for Science, Policy & Outcomes.

Model citation for this volume:

Lloyd, J., Nordhaus, T., Sarewitz, D., Trembath, A., eds. 2017. *The Rightful Place of Science: Climate Pragmatism*. Tempe, AZ: Consortium for Science, Policy & Outcomes.

ISBN: 069289795X

ISBN-13: 978-0692897959

FIRST EDITION, JUNE 2017

CONTENTS

CONTRIBUTORS

The following authors participated in the development, writing, and editing of the reports on which this book is based. The opinions expressed in this volume are those of the authors, and do not necessarily represent the views of the institutions with which the contributors are affiliated.

Mark Caine	African Union Commission
Netra Chhetri	School for the Future of Innovation in Society, Arizona State University
Gary Dirks	Global Institute of Sustainability, Arizona State University
Loren King	Institute for Public Policy Research
Frank Laird	Josef Korbel School of International Studies, University of Denver
Jason Lloyd	Consortium for Science, Policy & Outcomes, Arizona State University
Jessica Lovering	The Breakthrough Institute
Max Luke	New York University
Lisa Margonelli	Author of *Oil on the Brain: Petroleum's Long, Strange Trip to Your Tank*
Todd Moss	Center for Global Development, Center for Energy Studies at the Baker Institute, Rice University
Ted Nordhaus	The Breakthrough Institute

Roger Pielke Jr.	Center for Science and Technology Policy Research, University of Colorado at Boulder
Mikael Román	Office of Science and Innovation, Embassy of Sweden*
Joyashree Roy	Department of Economics and Global Change Programme, Jadavpur University
Daniel Sarewitz	Consortium for Science, Policy & Outcomes, Arizona State University
Michael Shellenberger	Environmental Progress
Kartikeya Singh	Center for Strategic and International Studies
Peter Teague	Philanthropic Consultant
Alex Trembath	The Breakthrough Institute

* Dr. Román participated in the Climate Pragmatism workshops and as contributor in his personal capacity only, and the statements in this book do not represent the views of the Ministry of Enterprise and Innovation or the Swedish Government.

FOREWORD

Current discussions on the role of energy production and use in the climate debate often revolve around two conflicting perspectives. One view is that energy production and use is a key source of carbon emissions that contribute to climate change. This view leads to recommendations for policy measures that seek to curtail the use of certain energy sources or promote energy conservation.

The other view is that countries that have limited access to modern energy resources are likely to rely on traditional sources in ways that lead to ecological degradation and possibly increased carbon emissions. This is the case, for example, in regions of the world that rely on biomass harvesting for energy production and use. By seeking to identify areas where energy production and use can be reduced or curtailed to address environmental concerns such as climate change, neither perspective offers a way for emerging countries to move away from inefficient traditional energy sources and effectively meet their citizens' growing energy demands.

As *The Rightful Place of Science: Climate Pragmatism* underscores, these views are based a false dichotomy between energy and environment. The fundamental role that energy plays in human well-being puts it at par with other critical resources, such as water. In fact, it can be argued that the right to energy underlies the very idea of human rights itself, given its critical role in human survival. This is evident when looking at emerging economies that are still grappling with meeting their basic needs. Much of the ecological destruction in regions such as Africa is a result of low levels of modern energy usage; efforts to curtail its expansion are detrimental to both human well-being and environmental integrity.

One way to decouple energy production and use from environmental degradation is to expand its abundance, diversify its sources, and accelerate innovations in clean energy technologies. Abundance and diversity provide users with the flexibility to meet their needs and expand opportunities for environmental management. Innovation—particularly collaborative innovation efforts that focus on a diversity of deployment contexts—support abundance and diversity by making energy production and consumption cleaner and cheaper. It is through expanded access to energy services that one can address various sustainable development goals and build prosperous, resilient societies.

Increased energy production and use is essential for fostering economic prosperity. It is through prosperity that societies can build up the capacity to address challenges such as poverty and build more environmentally resilient communities. To the contrary, societies that have limited access to energy are unlikely to improve their well-being and effectively reduce the ecological impacts of their activities. If we wish to foster communities that are adaptive and resilient to a range of challenges, including the consequences of a warmer climate, access to abundant and affordable clean energy is central to that effort.

The Rightful Place of Science: Climate Pragmatism stresses the importance of creativity and innovation in addressing environmental challenges and improving human well-being. This view requires a different approach to the role of energy in emerging economies. The challenge is not simply having the ability to power certain activities. It is appreciating the important role that energy-related technologies and infrastructure play in overall economic development. Energy infrastructure is a critical source of industrial development that fosters economic transformation. The diffusion of energy-related technologies and skills into the wider economy provides additional possibilities for economic diversification.

The development of the steam engine, for example, became a source of a wide range of complementary technologies that added to the economic transformation that many now-affluent countries experienced in the nineteenth and twentieth centuries. It was through seeking to improve the efficiency of steam engines that the second law of thermodynamics was formulated. And with its understanding came many other industrial applications that helped to improve human well-being. Today's advances in renewable and other low-carbon energy sources should also be viewed as technological revolutions, in addition to their role in powering economic activities.

Access to abundant energy and diverse sources is not just a right for emerging regions of the world, but a foundation for the improvement of human well-being, improved environmental management, and—as the steam engine example illustrates—continued technological innovation. *Climate Pragmatism* serves as a clarion call for humanity to abandon the misguided view that increased energy use is at odds with efforts to address climate change.

As this volume stresses, the answer lies in being able to appreciate the importance of energy as a driver of human creativity and innovation. It is not a constraint on the human capacity to act in the interest of improved climate mitigation and adaption.

Climate Pragmatism also cautions against the view that emerging regions of the world, especially Africa, can leapfrog into the age of low-impact, decentralized renewable energy sources that just meet their current needs. This view is reminiscent of the hobbling ideas about "appropriate technology" peddled in the 1970s. Off-grid solar energy sources are an important starting point for isolated applications, but they are a poor substitute for a more robust approach that includes local technological innovation in energy systems. Like in agriculture, a subsistence approach to energy production and use runs counter to efforts to promote prosperity, resilience, and human well-being.

A rights approach to energy argues against placing advance restrictions on the ability of emerging economies to access all adequate technologies available in a safe and responsible way. Such access will require considerable effort on the part of the international community to forge inclusive international energy partnerships and accelerate innovation in the energy sector.

By focusing on creativity and innovation in energy abundance and diversity, *Climate Pragmatism* has successfully brought fresh thinking on how to engage in a more positive discourse on the relations between energy, climate change, human development, and adaptation. This refreshing approach advocated in this book will help foster more productive international engagements and trust among industrialized and emerging nations and bring new voices to help address some of humanity's most pressing problems.

The power of the *Climate Pragmatism* vision lies in its ability to abandon long-held dogmatic views about the relationship between energy and climate change in favor of a well-founded faith in human ingenuity. It offers a pragmatic way forward that helps policymakers, researchers, and practitioners escape the false and stultifying dichotomies between economic growth and sustainability, between energy consumption and climate change, and between climate mitigation and adaptation.

Calestous Juma

Professor of the Practice of International Development
Belfer Center for Science and International Affairs
Harvard Kennedy School, Cambridge, MA, United States

INTRODUCTION

The Climate Pragmatism project is a partnership between the Consortium for Science, Policy & Outcomes (CSPO) at Arizona State University and the Breakthrough Institute in Oakland, CA. From 2013 to 2016, the project held three workshops that brought together experts from a range of disciplines to explore promising new approaches to the climate change challenge. The results of these meetings were originally published in three reports: Our High-Energy Planet *(2014),* High-Energy Innovation *(2014), and* Adaptation for a High-Energy Planet *(2016). This book is adapted from those reports.*

In this introduction, three of the project's central personnel discuss Climate Pragmatism's origins and goals. Ted Nordhaus is the Co-Founder and Executive Director of the Breakthrough Institute; Joyashree Roy is a Professor of Economics at Jadavpur University in Kolkata, India; and Daniel Sarewitz is the Co-Director of CSPO and Professor of Science and Society at the School for the Future of Innovation in Society at Arizona State University.

How did this project come about? What was the intellectual and political context for this work?

Ted Nordhaus: The intellectual foundation for the set of ideas presented in this project was first articulated, at least by us, in "The Hartwell Paper: A New Direction for Climate Policy After the Crash of 2009" (2010). In that paper, Dan Sarewitz, Roger Pielke Jr., myself, and a number of other colleagues outlined an alternative framework for both thinking about the climate challenge and addressing it. A year or so later, we at the Breakthrough Institute published a paper called "Climate Pragmatism: Innovation, Resilience, and No Regrets" with a number of U.S. coauthors. That report focused on how the framework could be applied in a U.S. context. Out of that work, there was a sense that we needed to further refine what a climate pragmatism framework looked like in the context of energy access, global energy innovation, and adaptation. That was really the driver: our recognition that the old framework was collapsing. Following the failure of efforts to create a follow-up to the Kyoto Protocol at the 2009 Copenhagen climate summit, there was an effort to put a new framework together. The idea behind the reports collected in this book was to identify what the key dimensions of the climate-human development interface really needed to look like to make effective progress on that framework. Similar ideas have finally begun to emerge at the international level in the 2015 Paris Agreement.

Daniel Sarewitz: Both in the case of the Breakthrough Institute and their pedigree, starting with "The Death of Environmentalism: Global Warming Politics in a Post-Environmental World" (2007), and work I'd been doing with Roger Pielke Jr. and Steve Rayner since the mid-1990s, each of us had been following a separate initial thread but recognized that the conventional approach to climate change wasn't going to work. It was insufficiently attentive not just to the political dynamics of climate change, but to the problems of global human development, equity, and distribution. We shared this sense that the world needed to address climate change, but it also had a bunch of other challenges that couldn't be separated from climate change. That is, you couldn't achieve a pragmatic political pathway to deal with climate if you didn't deal with these other challenges as part of that pragmatic pathway; this project marked an advance on that realization. I would also add that it was really hard to get money to fund this kind of work, but we were fortunate in that the Breakthrough Institute found a funder, Peter Teague at the Nathan Cummings Foundation, who was courageous enough to recognize the need to bail out of the mainstream funding pathway and do something different if we really wanted to chart a better way forward.

Joyashree Roy: My first involvement was with the Breakthrough Dialogue, the institute's annual conference. When I went to the Dialogue, I saw that the discussions were different than the then-mainstream climate change discourse, in which the politics were stalemated. The discourse had become a kind of contestation between developed and developing countries. Developing countries understood that they had to go forward with their development agenda, given that so many people required their basic needs to be fulfilled; but they perceived that the climate change discourse was working as a barrier to their progress. And they started seeing the climate agenda as creating a conflict between the historical development of rich countries versus the current and future advancement of developing countries; it became a blame game, and it was not heading anywhere. The question had become, "Do we develop or do we protect the climate?" It was not, "How can we move forward with human aspirations but also meet the climate challenge?" I felt that the whole discourse was not looking for or offering solutions. A single silver bullet is obviously not possible, but advances in whole suites of technologies, for example, could be diffused and deployed worldwide so that the necessary progress can happen with fewer emissions. Scientific assessments, such as those from the Intergovernmental Panel on Climate Change (IPCC), also mentioned using a portfolio of options in this way. But the political discourse was stuck in a different place and not fully considering those options. The Breakthrough Dialogue provided a novel, open platform that was dedicated to exploring all the possibilities—trying to see what really works now, and what can work in the future. Climate pragmatism, as discussed at the Dialogue, made sense to me because it could provide a way from contestation to a more constructive debate.

What does "climate pragmatism" mean?

Dan: The question follows perfectly from what Joyashree is saying. Where climate pragmatism starts is with the recognition that making action contingent on solving the political problem—that is, the sources of contestation both internationally and within nations that Joyashree mentioned—really misses a huge opportunity. There are ways to move forward that don't require you to solve the fundamental political and value disagreements that brought climate policy and politics to gridlock. There are tools out there, and ways of looking at the world, that really ought to enfranchise broad groups of formerly competing perspectives. First, the idea that the world needed a lot more energy to support the potential for people to reach their developmental aspirations, that this could not be held hostage to climate change debates, and that in fact, as we articulated it in the first of the Climate Pragmatism workshops, we needed a "high-energy planet"—all of that was a good thing, not a bad thing. And as we recognized and formalized in the second workshop, we know a lot about how to innovate and drive innovation. This innovation doesn't require a global consensus on climate change action: we can innovate in the energy space and move simultaneously towards a high-energy planet, but also a clean-energy planet. And then third, adaptation can no longer be the poor second cousin of the climate change action agenda. All societies need to be resilient in the face of a complex and sometimes harsh environment, regardless of future climate change, and climate pragmatism could be a front where interests converge around the need to ensure more resilient societies. So the idea of climate pragmatism is kind of obvious: let's do the things that provide broad benefits regardless of one's particular set of commitments in the climate debate. That's the position we're trying to advance in specific and in general in the book, and I think it's one that the world is beginning to catch up with.

How does climate pragmatism differ from other approaches to climate change? Is there any overlap between the policy ideas in this book and the approach embodied by, for example, the 2015 Paris Agreement?

Joyashree: The Paris Agreement is actually talking about solutions, which marks a change that the climate pragmatism ideas also tried to flag and initiate. Climate pragmatism focuses on the fact that, historically, humanity has done many good things in terms of addressing environmental problems and advancing human development goals, and seeks ways that those positive actions can be scaled up. Previously, the climate debate was stalled by the idea that humanity has done almost everything bad: we have created technologies which are not solving the problem, we have invented things which created this climate problem in the first place—it was a more problem-centric focus, an argument that humanity has done all the wrong things in the past. We had to abandon that tired thinking, to start fresh in order to move ahead. Climate pragmatism offered that alternative trajectory, acknowledged the need for a high-energy planet, and brought a positive tone to the whole stalemated discourse. With a focus on what's possible in the near- and medium-term, it produces a solution-specific dialog that decision- and policymakers can build on in pursuing climate actions like the Paris Agreement, with positive long-term implications.

Ted: In terms of the arguments that are laid out in this project and this book, we suggest that there are a number of critical boundary conditions that successful action to address anthropogenic climate change has to meet. The first is that human development needs must be met globally, and that very large populations of people will need to make the transition fully to something that approaches what we in the developed world consider to be modern living standards. The second, which follows from that, is that energy plays a critical role in that process. *Access* to energy simply isn't sufficient; what's required are modern levels of energy consumption. Once you establish that, two things become apparent. The first is that societies will become more resilient to whatever level of climate change we experience in significant part as a function of living in high-energy societies capable of providing high living standards. In other words, resilience is a co-benefit of achieving those higher living standards and higher levels of energy consumption. And the second thing is that advancing innovation of energy systems at quite significate rates will be necessary if those boundaries conditions are to be met at the same time that climate change—to the extent possible—is going to be mitigated. Both climate mitigation and climate resilience are essentially co-benefits of higher energy consumption and energy innovation. And I think when you look at the shift that happens and is reflected in Paris, we move from this idea of putting legally binding emissions reduction commitments at the center of the international effort to address the issue, and move into the Intended Nationally Determined Contributions (INDC) framework, where basically nations make pledges around what they are going to do to decarbonize their energy systems. Those are reflections of national development priorities and geopolitical priorities for nation states—which is where the opportunities for decarbonization actually are. So in the Paris Agreement, you see a reframing of the issue that really starts with work that folks like Steve Rayner and Dan and Roger did going back many years. That work was then picked up in "The Death of Environmentalism," which says that if you're going to make progress on these global environment problems, you need to take the environment out of the center of your politics and put human development at the center of it. That is all summarized in the Hartwell Paper and then delineated with a lot more specificity in this book, pulling together many years of work in a fairly explicit and extensive way.

What is the vision for a world that pragmatically addresses climate change?

Dan: I'll make the point that we made in the Hartwell Paper: a pragmatic approach to addressing climate change is one that doesn't obsessively focus on climate change. It's one that recognizes that climate change is actually an emergent outcome of many different types of complex human activities. I appreciate Joyashree's point about how one aspect of climate pragmatism is a move from negativity and doom-saying to a solutions orientation and a search for opportunity. The vision for a world that pragmatically addresses climate change is one that isn't constantly insisting that we're on the verge of disaster, but rather recognizes the risks in climate change and incorporates those into a whole range of activities that humans need to be taking seriously anyway. We need clean energy, we need cheap energy, we need energy security, we need everyone to have access to the amount of energy necessary to fulfill their aspirations, we need people to live in environments that are resilient in the face of a capricious human-nature relationship. In a way, the vision for the world that pragmatically addresses climate change is one that seeks opportunities to advance its core values and in doing so, just happens to address the challenge of climate change too. The climate pragmatism argument is that those two things are actually integrally connected, that it is in the process of meeting human needs and aspirations that climate change will be addressed—as Ted said, it's a co-benefit. So part of this vision is to accept that the focus is less on climate change and more about human dignity and development.

Joyashree: It's also important to say why we call it "pragmatism," because that's how this discourse puts the dialog on a different track. For the earlier mainstream climate change conversation, human action was considered a barrier for progress. Climate pragmatism starts with the premise that human beings are actually capable of finding solutions. Regaining faith in human capability, capacity, and ingenuity is something valuable that climate pragmatism has brought into the discussion. Because otherwise the whole discourse is that human actions are wrong and tend to produce bad outcomes. I want to highlight the fact that climate pragmatism has reinstated faith and confidence in humanity's ability to address our problems—a very important contribution when attempting to confront such an enormous challenge.

Dan: I love that. Just to amplify that point, I think that in some of the alternative views of climate change, there's a real suspicion of technology and suspicion about visions of human progress. Joyashree, you were saying—and I think this is right—that one may or may not hold those suspicions, but there's no way forward without wise human action and without immense support from new technologies. Climate pragmatism represents an optimistic view of human potential, and that makes it different from the standard ways of talking about climate change.

Joyashree: And if you look at the Paris Agreement, that theme is now recognized, as Ted mentioned. In an increasingly fragmented world, a positive vision like climate pragmatism has the potential to drive the discourse from contestation to cooperation, to national autonomy in decision making, and mass action to deliver a global good.

What do you hope the publication of *The Rightful Place of Science: Climate Pragmatism* accomplishes?

Ted: I don't think I can speak for all of the authors, or even all the editors, but I think we reached a moment a while ago—one that still influences a lot of the public arguments—where the old framework around which the climate debate hinged is basically exhausted. You still have various parties arguing, on the one side, for really dramatic emissions cuts—organizations like 350.org focused on keeping climate change under 1.5 degrees Celsius—and on the other side, the Trump presidency and a set of folks on the right claiming that this is all a gigantic conspiracy. In the midst of that, the world mostly has continued on a business-as-usual trajectory for decades now. When we think about where this points us going forward, it is to start to get our heads around—and Dan alluded to this already—that global emissions are going to go well past 450 parts per million of atmospheric carbon dioxide concentration, and if most projections are right, global temperatures will increase past 2 degrees Celsius. We need to be thinking about climate policy on both the mitigation side and the adaptation side in a world of 3 degrees or more of warming, which—even if we're substantially more successful at mitigating than we have been to date—is the likely trajectory. Understanding that the world wasn't saved at emissions levels under 2 degrees and isn't lost beyond 2 degrees is an important starting point for thinking about what's practical in terms of the continuing evolution of global energy systems and the ways in which development, mitigation, and adaptation strategies need to be much better integrated. We're only beginning to think about what that looks like and consider what's possible, to come to terms with both the uncertainties in our understanding of the climate and some of the implications of those uncertainties. And at a moment where a lot of the longstanding international institutions and norms of the post-World War II era are being questioned, we are just starting to develop an international framework that might succeed at mitigating the worst climate change impacts, adapting to what can't be avoided, and doing that all in a context where nine or ten billion people have an opportunity to live modern lives with physical and economic security and human dignity.

Dan: Part of the message that we want to propagate, to reiterate Joyashree's point a bit, is that "We can do this." The world's not a perfect place and the unexpected will happen, we can't manage and control an incredibly complex future, but there are core aspects of the climate challenge that we know how to deal with based on our experience of doing similar things in the past. We know how to rapidly accelerate innovation in areas of energy and other important technologies—we've done that before. We know how to protect societies both affluent and poor from the ravages of natural hazards. The Dutch have lived below sea level for centuries. More recently, Bangladesh has done an incredible job of reducing the human toll of flooding and typhoons. There are therefore proven human pathways here, and we know how to get on those. What we need to do is stop focusing on the ideology and politics of one particular pathway, and recognize there are many ways forward; the climate problem actually presents an opportunity for achieving human aspirations across many dimensions. Without at all minimizing the challenges that climate change provides, what climate pragmatism does is provide a hopeful way of understanding how human ingenuity has in the past and can in the future deal with this problem.

Joyashree: What needs to be emphasized a little more is that the scientific understanding of the climate issue, including pragmatic ways of dealing with it, is progressing faster than institutions can act or change. An important open question—potentially a fruitful line of future research—is how to make progress more quickly, even if the institutions we rely on to structure and advance that progress don't change that quickly. In terms of technological innovation in the clean energy sector, for example, we have a lot to learn from the private sector regarding development and diffusion. Otherwise it becomes an excuse: "The policy institutions aren't innovating, the constraints are not changing, so addressing climate change is not happening." Different fields of science and technology and business can help in our search for more practical solutions and in implementing them quickly. That type of bridging of knowledge really needs to be done, and that's something I think climate pragmatism is well positioned to do.

OUR HIGH-ENERGY PLANET

— A CLIMATE PRAGMATISM PROJECT —

By Mark Caine, Jason Lloyd, Max Luke, Lisa Margonelli, Todd Moss, Ted Nordhaus, Roger Pielke Jr., Mikael Román, Joyashree Roy, Daniel Sarewitz, Michael Shellenberger, Kartikeya Singh, and Alex Trembath

APRIL 2014

1

ENERGY FOR DEVELOPMENT

Access to affordable and reliable energy is a prerequisite for human development. Modern energy in the form of electricity and liquid fuels undergirds every aspect of contemporary life, from education to healthcare, manufacturing to telecommunications, agriculture to transportation. Affluent countries have spent two centuries unlocking the potential of widespread energy access and consumption to improve the lives of their citizens and build dynamic, prosperous societies. Today, hundreds of millions of people in the developing world are using modern energy to escape poverty.[1] So overwhelming and undeniable is the importance of energy to quality of life that any agenda intent on advancing human development and dignity must place universal and equitable access to modern energy at its center.

Rapidly growing countries like Vietnam, Brazil, India, and South Africa have accelerated development and improved citizens' lives by focusing on economic growth, industrial productivity, and energy system modernization. As the poorest countries of the global South transition from agrarian to industrial societies, they are likely to follow a similar path. They will require much more ener-

[1] United Nations Development Programme, *Human Development Report 2013, The Rise of the South: Human Progress in a Diverse World* (New York, NY: UNDP, 2013).

gy as they do so. In seeking to assist these least-developed countries in their modernization, international organizations such as the United Nations have highlighted the role of energy access in achieving development objectives. This is commendable. But the energy access proposals outlined by these groups are constrained by a simultaneous focus on minimizing greenhouse gas (GHG) emissions in an attempt to deal with climate change. As a result, the levels of energy access envisioned by these initiatives are often inadequate for driving long-term socioeconomic development.

Energy equity means that all people have access to the kind and quantity of energy that allows them to achieve their development aspirations. This entails the availability, reliability, and affordability of a diverse range of energy services. Energy services include heat, illumination, cooling, communication, and mechanical power.

Rather than limiting energy access and consumption on the basis of their potential climate change impacts, a coherent strategy for human development begins with the assumption that energy equity is necessary for a just, prosperous, and environmentally sustainable society. By building out the worldwide energy system in support of human dignity and widely shared prosperity, we create fertile conditions for the emergence and scaling of new innovations that will generate progressively lower-carbon developmental pathways. It is precisely the massive expansion—rather than contraction—of energy systems, carried out primarily in the dense population centers of the global South, that provides the context and opportunity for a robust, coherent, and ethical response to the challenges we face.

Our high-energy framework is prescriptive only in its commitment to energy equity. It is about creating options for societies to grow and prosper as they wish. We do not advocate any particular economic or social arrangement, energy system configuration, or vision of society in relation to nature. These are things that communities and societies must determine on their own; access to modern energy makes that choice possible. The appropriateness of energy technologies is determined by their ability to meet the current and—in order to avoid locking in energy poverty through low levels of access—future energy needs of individuals and communities.

Our approach combines a commitment to pragmatism—a clear-eyed focus on what works in practice, rather than what's ideologically acceptable—with an insistence that all humans deserve access to sufficient energy services to achieve the quality of life currently enjoyed by people in economically developed regions of the world. A high-energy planet with universal access to affordable, cleaner, and plentiful energy is the most practical way to secure this socioeconomic development while ensuring environmental protection.

2

WHAT IS ENERGY ACCESS?

Energy, Productivity, and Human Development

Economic productivity and social well-being have co-evolved over the past two centuries as central features of human development. This process is characterized by sustained growth in the capacity to cultivate knowledge and skills that lead to further innovation, prosperity, and resilience. Central to this co-evolution has been the ability of citizens and businesses to take advantage of affordable and reliable energy, which today comes mainly in the form of electricity and liquid fuels like gasoline.

The relationship between access to modern energy and quality of life is well established. Two hundred years ago, the economies of Western Europe and the United States began to shift to industrial modes of production. Their populations began the transition from traditional biomass fuels to the precursors of the energy sources in use today, and living standards improved dramatically. The same is true now in the developing world. As poor people gain access to electricity and cleaner fuels, they typically enjoy longer, healthier, and more prosperous lives.

Widespread energy access also powers growing economies. The transformation of natural energy assets into usable energy services allows not just for household lighting and electricity, but also modern infrastructures and industrial practices that have positive social impacts. Affordable energy is used to drive tractors, create fertilizers, and power irrigation pumps, all of which improve agricultural yields and raise income. Inexpensive and reliable grid electricity allows factory owners to increase output and hire more workers. Electricity allows hospitals to refrigerate lifesaving vaccines and power medical equipment. It liberates children and women from manual labor and provides light, heat, and ventilation for the schools that educate the workforce.

While there is no single or linear path to a modern energy system, there is a pattern common to many societies. As countries like the United States and Great Britain shifted from agrarian to industrial to postindustrial societies, they developed more efficient, flexible, and convenient energy sources to power increasingly complex economic activities. From an almost total reliance on biomass fuels like wood and charcoal in the early 1800s, advanced economies now depend on reliable, grid-based access to a diverse mix of energy sources. These include coal, oil, natural gas, hydropower, nuclear, and renewables like wind and solar. Technical innovation, economies of scale, government investments, and competitive markets for energy services improved the performance of these energy systems, lowered their costs, enhanced the services they provide, and spurred the generation of new services that benefitted lives and livelihoods.[1]

In this sense, we understand the development that energy supports to mean much more than just economic growth. Rather, human development involves building knowledge and skills that allow a society to innovate, solve problems, enhance productivity, and improve pro-

---

[1] Vaclav Smil, *Energy Transitions: History, Requirements, Prospects* (Santa Barbara, CA: Praeger, 2010).

cesses, capabilities, and technologies.[2] Historically, such increases in society-wide capacity have allowed countries and their citizens to reduce the amount of carbon they burn per unit of energy produced—that is, to decarbonize their energy systems. Development advances societies along trajectories of their choosing, as people with access to energy and the opportunities it enables acquire freedom of choice about how they live and what livelihoods they pursue. Societies that are able to meet their energy needs become wealthier, more resilient, and better able to navigate social and environmental hazards like climate change. Therefore, like freedom from violence, hunger, and the diseases of poverty, access to sufficient levels of energy must be understood as a cornerstone of human development.

While the world has made considerable progress in expanding energy provision, billions of people still lack access to sufficient modern energy. Today, the poorest three-quarters of the world's population use just one-tenth of the world's energy. Over one billion people around the world—five hundred million of them in sub-Saharan Africa alone—lack access to electricity. Nearly three billion people cook over open fires fueled by wood, dung, coal, or charcoal.[3] The health consequences of these energy use patterns are severe: every year, indoor air pollution causes two million premature deaths, one million cases of chron-

---

[2] Rob Byrne, Adrian Smith, Jim Watson, and David Ockwell, *Energy Pathways in Low-Carbon Development: From Technology Transfer to Socio-Technical Transformation* (Brighton, UK: STEPS Centre, STEPS Working Paper 46, 2011).

[3] International Energy Agency, "Measuring Progress Towards Energy for All," Ch. 18 in *World Energy Outlook 2012* (Paris, France: OECD/IEA, 2012).

ic obstructive pulmonary disease, and half of all of pneumonia deaths among children under the age of five.[4]

An Urban Planet: People to the Power

Urbanization is shaping the context for expanded energy access and energy innovation. Demographic trends in the developing world demonstrate an inexorable transition from rural, pastoral societies toward urban, industrial societies.[5] This transition makes countries wealthier, more productive, and more innovative, but sustaining these advances requires large quantities of energy. High population densities require large-scale, centralized energy sources. Low-carbon options for these sources include hydropower, coal with carbon capture and sequestration (CCS), advanced nuclear, and natural gas, along with efficient networked configurations of gas pipelines and electricity grids.[6]

Historically, most rural dwellers have gained access to modern energy services by moving to cities.[7] To be sure, rural electrification efforts have played an important role

---

[4] Kirk R. Smith, Sumi Mehta, and Mirjam Maeusezahl-Feuz, "Indoor air pollution from household use of solid fuel," in *Comparative Quantification of Health Risks: Global and Regional Burden of Disease Attributable to Selected Major Risk Factors*, Vol. 2, Majid Ezzati, Alan D. Lopez, Anthony Rodgers, and Christopher J. L. Murray, eds. (Geneva: World Health Organization, 2004).

[5] United Nations Human Settlements Programme, *Planning Sustainable Cities: Global Report on Human Settlements 2009* (London, UK and Sterling, VA: UN Habitat and Earthscan, 2009).

[6] See Jesse H. Ausubel, "Decarbonization: The Next 100 Years," 50th Anniversary Symposium of the Geology Foundation, Jackson School of Geosciences, University of Texas (Apr. 25, 2003).

[7] Douglas Barnes (ed.), *The Challenge of Rural Electrification: Strategies for Developing Countries* (Washington, DC: Resources for the Future, 2007).

in ensuring access in industrial countries. But such success in electrification has been facilitated at least in part by the fact that the rural population in need of electricity was shrinking as expanding urban industrial centers drove economic growth. There remain, of course, billions of people who live in rural communities, and there will be for many decades to come. The solutions to energy poverty in remote, rural villages will, of necessity, look quite different from solutions in urban areas. But most of the increase in global energy access through the next century is likely to be the result of urbanization combined with better urban electrification efforts.

Urbanization contributes to universal electrification in three distinct ways. First, it shifts populations into denser living configurations that are less costly to electrify. This explains why, for instance, 69 percent of urban citizens in sub-Saharan Africa have electricity access whereas only 18 percent of rural residents do.[8] Second, urbanization reduces the size of rural populations that must be served through expensive grid extensions. And third, urban societies are more affluent and better able to provide a critical mass of consumers who can support a financially sustainable energy system.[9] These factors underlie the historical pattern of universal electrification in rich countries and the provision of energy services to their rural or remote populations.

Nonetheless, the archetypal image evoked by efforts to address energy poverty is the desperately poor rural village, not the sprawling informal urban slum. And while

---

[8] Worldwide, urban and rural residents have electricity access at rates of 96.4 percent and 73 percent, respectively. More details can be found at: http://data.worldbank.org/indicator/EG.ELC.ACCS.UR.ZS

[9] World Bank, *Addressing the Electricity Access Gap* (Washington, DC: World Bank Group, June 2010).

the critical needs of rural communities must not be ignored, global energy and climate policies, in order to succeed, will need to focus much more heavily on the needs of rapidly growing urban populations in developing economies. National electrification plans that envision generating capacity and infrastructure to power growing cities and densely populated peri-urban regions should be priorities for development organizations and multilateral institutions like the United Nations.

Energy Access as a Public Good

The kind of large-scale, long-term planning required for equitable energy access points to an important facet of the policy framework for achieving a high-energy planet: the need for effective and responsive governance that takes into account factors like the wider economy, environmental impacts, and public safety. Public cooperation with the private sector is the best way to achieve this governance. The provision of reliable energy services is in many ways unique—enormously complex, operating over decades, risky for investors, and with weak market signals. The private sector alone lacks the capacity and incentives to make energy systems significantly cleaner or more equitable, especially for very poor and marginalized energy consumers.

As a result, energy modernization has been driven, almost everywhere, by a strong public commitment to expanding access to modern energy services in pursuit of broader economic development goals. Public utility companies or highly regulated monopolies sought secure energy supplies and low-risk efficiencies to ensure affordable prices, guaranteed profits, and high rates of economic growth.[10] States partly subsidized the creation

[10] Smil, *Energy Transitions.*

and extension of electric grids, typically through loan-interest financing, early losses recovered through increased consumption later, and strategic tariff structuring. Individual and commercial consumers then came to expect the most cost-effective and reliable energy sources to power their homes, vehicles, and businesses.

The Sabarmati Thermal Power Station provides electricity to the cities of Ahmedabad and Gandhinigar, India. *(Photo credit: Koshy Koshy.)*

Guaranteeing access to modern energy services is a way of utilizing collective national resources to create a public good from which all people benefit. Effective states have long recognized this, which is why the energy infrastructure of all industrially developed nations—and those of later-industrializing countries like China, Chile, Mexico, Thailand, Tunisia, and others—has been dominated by either public or heavily regulated institutions.[11] Just as no country leaves the construction and maintenance of its transportation networks or public health systems to the

[11] Liberalized (or deregulated) electricity markets are a relatively new phenomenon in wealthy countries, and came about only after near universal access had been achieved.

complete mercy of the free market, public policy and financing for building and modernizing a country's energy system—one of the most basic responsibilities of a functioning state—is necessary. Only public sector involvement can devote resources on the requisite scale, ensure that abundant and affordable energy services reach all citizens, and create the stable economic and political conditions necessary to expand private investment.

Countries as varied as Brazil, Indonesia, and Vietnam have made substantial progress toward universal energy access. Through public financing and proactive legislative support, they have created conditions for private investment in the energy system, fostered competition and efficiency in end-use technologies, and mitigated market failures like underinvestment in innovation, infrastructure, and rural electrification. Indeed, countries that have successfully electrified have relied on encouraging productive use of electricity like water pumps for agriculture, or refrigeration for food and medicine distributors.[12] Such public leadership in the energy sector—often demanded by industrial sectors that need reliable energy in order to be competitive—typically instigates a virtuous cycle of increased energy access, rising incomes and political agency, and more responsive policies that provide citizens with plentiful, reliable, and cleaner energy services.

---

[12] Barnes, *The Challenge of Rural Electrification.*

3

THE ENERGY-CLIMATE CHALLENGE

Decarbonization

Creating a high-energy planet and building out the infrastructure necessary for universal and equitable access to energy involves immense changes in technical, social, economic, and environmental systems. These changes often have multiple benefits. Transitioning from dung and wood to coal, natural gas, nuclear, and other advanced energy sources alleviates many problems caused by a lack of energy access or the use of low-grade energy sources. These problems include deforestation, soil erosion, black carbon emissions, and respiratory ailments that lead to premature death.

Modern energy is crucial not just for human development, but also for environmental quality, as it allows developing nations to move away from traditional, polluting energy sources and toward cleaner fuels. Modern energy systems use relatively less carbon per unit of energy produced than pre-industrial systems. Charcoal, for example, which is a biomass fuel still used in many parts of the developing world, is almost pure carbon. Over time, consumption of hydrocarbon fuels has shifted towards sources of energy with ever-higher hydrogen-to-carbon ratios. Natural gas has half the carbon content of coal, and nuclear power and renewable sources emit no carbon dioxide in energy production.

This historical path of decarbonization has not prevented a continual rise of global carbon emissions because more efficient, affordable energy services have precipitated dramatic increases in energy consumption by growing populations. Much of the energy that underpins worldwide economic activity comes from burning fossil fuels that emit greenhouse gases and other pollutants, making the global energy system the largest contributor to global climate change. This fact has created the appearance of a conflict between the energy needs of developing countries and the need to address climate change. The logic of this conflict suggests that because energy is implicated in the climate change problem, energy production and consumption should be minimized and emissions reduced whenever and wherever possible.

This idea that energy use must be constrained, or restricted to specific zero-carbon technologies, informs the dominant framework within which energy and human development are understood by many environmental organizations, donor governments, and multilateral development agencies, typically based in the United States and Europe.[1] It is often manifested in subtle and implicit ways. This framing is especially evident in the low quantitative thresholds of international energy access initiatives;[2] in "climate stabilization scenarios" and energy consumption

---

[1] For example, the Brundtland Report, the foundational document of sustainable development that we will discuss more in Chapter 6, concludes: "It is clear that a low energy path is the best way towards a sustainable future." *World Commission on Environment and Development: Our Common Future* (Oxford, UK: Oxford University Press, 1987).

[2] See, for example: World Bank, *Global Tracking Framework* (Washington, DC: World Bank Sustainability for All Program, 2013).

projections that assume billions of people will remain without access to modern energy for the foreseeable future;[3] and in international climate negotiations that focus on low-emission hardware transfers rather than social and economic development along technological pathways that do not trade near-term energy access for long-term climate goals.[4]

The energy access report from the UN's Advisory Group on Energy and Climate Change (AGECC), for example, notes that "energy facilitates social and economic development," while arguing that the world's least-developed countries must "expand access to modern energy services ... in a way that is economically viable, sustainable, affordable and efficient, and *that releases the least amount of GHGs*."[5] In other words, the United Nations supports human development through expanded energy access, so long as that energy does not come from the cheap fossil fuels that wealthy, developed countries spent the past two centuries burning, and that in many developing world contexts remain the cheapest, most reliable, and most versatile energy options.

Energy access initiatives are therefore often structured within a dominant climate narrative that denigrates—

---

[3] Roger Pielke Jr., "How Much Energy Does the World Need? Clarifying the 21st Century Energy and Climate Challenges," The Breakthrough Institute (26 Feb. 2013).

[4] Rob Byrne, Adrian Smith, Jim Watson, and David Ockwell, *Energy Pathways in Low-Carbon Development: From Technology Transfer to Socio-Technical Transformation* (Brighton, UK: STEPS Centre, STEPS Working Paper 46, 2011).

[5] Advisory Group on Energy and Climate Change (AGECC), *Energy for a Sustainable Future: Summary Report and Recommendations* (New York, NY: United Nations, April 2010), p. 8 (emphasis added).

often in explicitly moral terms[6]—many forms of energy production and consumption. In practical terms, such approaches seek to minimize, or even forbid, the use of the very energy technologies that enabled the prosperity of developed countries, and that are now spurring the rapid growth of many economies in the developing world.

The unacknowledged problem is that aside from nuclear and hydroelectric power, which most people in the environmental and development communities do not strongly advocate, there are not yet reliable, scalable, clean, base-load energy sources available to take the place of incumbent fossil fuels. Building out efficient grid infrastructure and modernizing the energy sector—including shifting from coal power to coal and natural gas with CCS, hydroelectric, and advanced nuclear power—are key processes in shifting to a low-carbon global energy system. Accelerating such a transition should be the priority of the climate and development communities, since the risks presented by climate change should not and will not be managed through limiting access to energy by the populations who need it most in order to thrive, innovate, and prosper.

Inadequate Energy Access Goals

The last few years have seen a growing international commitment to universal energy access. Former United Nations Secretary General Ban Ki-Moon made universal electrification one of his highest priorities, and the United Nations declared 2012 the Year of Sustainable Energy for All (SE4All). In June 2013, President Barack Obama an-

---

[6] See, for example: Friends of the Earth International, *Good Energy Bad Energy: Transforming Our Energy System for People and the Planet* (Amsterdam, Netherlands: Friends of the Earth International, 2013); and James Hansen, open letter to Michelle and Barack Obama (Dec. 29, 2008).

nounced a $7 billion effort to "Power Africa," aimed at extending electricity to 20 million households. In early 2016, the U.S. Congress passed "The Electrify Africa Act" (H.R. 2847) to extend electricity to 50 million people. These efforts involve the Export-Import Bank, the Overseas Private Investment Corporation (OPIC), the Millennium Challenge Corporation, the U.S. Trade and Development Agency, and the U.S. Agency for International Development. Power Africa also includes $9 billion in commitments from private companies such as General Electric.

These new initiatives have prioritized improved energy access as a means of achieving larger development objectives like economic growth, public health, and education. But, bounded by the conventional GHG-reduction framework outlined above, even when universal energy access is a declared objective, the thresholds are typically unacceptably low and far from equitable. The SE4All initiative, for example, claims that "basic human needs" can be met with enough electricity to power a fan, a couple of light bulbs, and a radio for five hours a day.[7] This is a baseline that someone from a rich country would not recognize as access at all. The average European consumes SE4All's yearly energy access threshold in less than a month,[8] and a typical American burns through that much electricity in a little over a week.[9] More importantly,

---

[7] AGECC, *Energy for a Sustainable Future.*

[8] Average per capita electricity consumption in Europe is approximately 1,500 kWh. See: World Energy Council, *Energy Efficiency Policies Around the World: Review and Evaluation* (London, UK: World Energy Council, 2008), p. 30.

[9] Eight point three days, to be exact. Average surveyed household electricity consumption is 11,280 kWh per year and average household size is 2.6 people, meaning per capita domestic consumption in the U.S. is 4,338 kWh per year. See: U.S. Energy Information Agency, "How much electricity does an American

such a baseline does not capture electricity used outside the home to power hospitals, schools, government facilities, commercial buildings, factories, and all the other places and activities that require electricity.[10]

Figure 1: Global Per Capita Electricity Consumption (kWh/year)

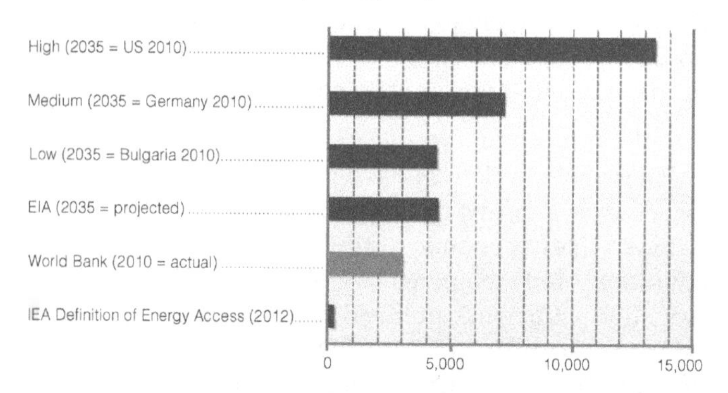

The same is true for other major energy access initiatives. The International Energy Agency (IEA) defines "energy access" as 500 kilowatt-hours (kWh) per year, or 100 kWh per person,[11] which is about 0.5 percent of the levels consumed by the average American or Swede, or 1.7 percent of the average Bulgarian. The World Bank's highest tier for energy access is less than 10 percent what the average Bulgarian uses. While acknowledging that these are initial targets and that efforts must be realistic in their

---

home use?" (2013), available at:
http://www.eia.gov/tools/faqs/faq.cfm?id=97&t=3

[10] Per capita electricity usage in the United States rises to nearly 14,000 kWh per year when total electricity consumption is incorporated. See World Bank, *World Development Indicators: Power and Communications* (2013), available at:
http://wdi.worldbank.org/table/5.11

[11] International Energy Agency, *World Energy Outlook 2010* (Paris, France: OECD/IEA, 2010), p. 249.

goals to be seen as credible, it is nonsensical to argue, as these goals implicitly do, that a household has achieved equitable access to modern energy when consuming 50 to 100 kWh per person annually—less than the average American's cable television box.[12]

The problem is not simply that these modest thresholds for energy access are low in comparison with high-income countries. After all, the immediate energy needs of poor communities in developing nations are much different from the energy demands of citizens in Canada or Germany. What *is* problematic is that these minimal targets can be met with energy technologies that have little capacity for scaling up and meeting the expanding needs of economically productive, non-household activities like manufacturing, transportation, or commercial agriculture.

Individuals, communities, and private enterprises thus lack a viable mechanism to move up the "energy ladder" for improved quality of life and greater productivity. Achieving negligible access thresholds with technologies like rooftop solar panels or cleaner cookstoves—rather than, for example, reliable grid connections—leaves other human development goals far out of reach. And because access initiatives are not typically part of long-term electrification strategies at the national or regional level, they do not support either effective development planning or the delivery of high-quality energy services to critical sectors of the economy.[13]

Put simply, initiatives like SE4All can fully succeed on their own terms without any meaningful or sustained de-

---

[12] See: Energy Information Agency, "End-Use Consumption of Electricity 2001," available at: http://www.eia.gov/emeu/recs/recs2001/enduse2001/enduse2001.html

[13] Morgan Bazilian and Roger Pielke Jr., "Making Energy Access Meaningful," *Issues in Science and Technology* (Summer 2013), pp. 74-78.

velopment on the part of energy-poor communities. Whatever the short-term benefit, a narrow focus on household energy and the advocacy of small-scale energy sources like solar home systems can, in fact, make it *more* difficult to meet the soaring increase in energy demand associated with moving out of extreme poverty.[14] Not only do such low thresholds drastically underestimate the magnitude of the energy access challenge, they also further entrench global inequities, distract financial and political capital away from more productive investments, and prevent people and nations from pursuing development paths that offer greater hope for reconciling their socioeconomic and environmental aspirations over the long term.

As a result, contemporary advocacy of sustainable energy expansion too often offers wildly inappropriate solutions, sometimes lifted wholesale from developed-world contexts that make no sense for energy-poor nations. For instance, despite being the world's sixth largest oil exporter, with vast reserves of natural gas, coal, and renewable energy, Nigeria has some of the lowest rates of energy access in the world.[15] The United Nations Development Programme's remarkable response to this situation is a project to "improve the energy efficiency of a series of

---

[14] A World Bank report notes "an important advantage of the grid connection over the [solar home system (SHS)], namely that increased electricity consumption by a grid consumer reduces its levelized cost per kWh significantly, whereas an increase in demand by a household relying on a SHS would require purchasing a second SHS, which would not reduce the levelized cost at all, or a larger unit, which would reduce the cost only slightly." See: World Bank, *Addressing the Electricity Access Gap* (Washington, DC: World Bank Group, June 2010), p. 36.

[15] Morgan Bazilian, Ambuj Sagar, Reid Detchon, and Kandeh Yumkella, "More heat and light," *Energy Policy*, Vol. 38 (2010), pp. 5409-5412.

end-use equipment ... in residential and public buildings in Nigeria through the introduction of appropriate energy efficiency policies and measures." Much of this $10 million project involves distributing compact fluorescent light bulbs.[16] The point is not that improving efficiency is ineffective in reducing carbon emissions, or that energy systems should not be made as efficient as possible in order to stimulate energy use for productive ends. It is that efficiency initiatives like these are entirely inadequate for development needs in energy-destitute countries like Nigeria.

The Nigeria case typifies why many present-day international sustainable energy initiatives, constrained by a framework that prioritizes emissions reductions even for the poorest, least-emitting countries, are ill suited to the development priorities of emerging economies. If universal energy access, properly understood as a broad development imperative, is to build on past lessons from the rich world and resonate with the ambitions of late-industrializing countries, it will in most cases entail the provision of reliable grid electricity.[17]

As with all large-scale technological transitions, this will be an evolutionary process, one that may begin with technologies like regional micro-grids. But it will require the committed support of development institutions to sustain and advance this process. On-demand grid electricity capable of powering commercial agriculture, modern factories, and megacities in the developing world will drive

---

[16] United Nations Development Programme, "Promoting Energy Efficiency in Residential and Public Sector in Nigeria," UNDP Project Document (2011), available at: http://www.ng.undp.org/energy/EE-project-document.pdf

[17] Douglas Barnes and Gerald Foley, *Rural Electrification in the Developing World: A Summary of Lessons from Successful Programs* (Washington, DC: UNDP/ World Bank Energy Sector Management Assistance Programme, Dec. 2004).

energy and development strategies for the foreseeable future. International energy access and climate initiatives that fail to align themselves with these priorities are largely irrelevant to the pursuit of decent living standards among those who currently lack modern energy access.

Energy Sector Build-Out

Energy systems grow as countries seek to meet their economic and social aspirations. Between 1990 and 2010, emerging lower-middle-income nations increased the percentage of their populations with access to electricity by 19 percent, to more than three-quarters of their populations. Regionally, the countries of southern Asia improved electricity access by 23 percent over the same time period, and North Africa went from 85 percent electricity access to almost 100 percent. Individual countries have seen even more impressive energy sector expansion: in Indonesia, for example, 94 percent of citizens now have electricity access, up from 67 percent in 1990.[18]

This expansion generates performance efficiencies and cost reductions that are advanced incrementally through technological demonstration, deployment, and improvement through day-to-day operations. These improvements within a stable system of institutions, skills, markets, political interests, and cultural forces can lead to what in hindsight are seen as breakthrough innovations toward new and cleaner energy technologies.[19] Concerted public and private efforts to develop and diffuse those innovations have succeeded in accelerating the rate at

---

[18] World Bank, "Energy Access," *Global Tracking Framework*, (Washington, DC: World Bank Group, 2013), pp. 262-270.

[19] Frank W. Geels, "Technological transitions as evolutionary reconfiguration processes: a multi-level perspective and a case-study," *Research Policy*, Vol. 31 (2002), pp. 1257-1274.

which energy systems shift to lower-carbon, more affordable technologies.

Innovations in unconventional natural gas production over the past decade, for example, lowered natural gas prices by 65 percent from 2009 to 2014[20] and led to historic reductions in U.S. carbon emissions.[21] In the 1970s and 80s, improved operating practices across the U.S. nuclear reactor fleet led to increases in capacity factors (the fraction of the actual-to-available power being generated) from an average of about 50 percent to 90 percent. This was a consequence of improved operating practices that came from use and experience, not from new technologies.

In other words, much of the opportunity for innovation in energy technologies over the next century will occur where and when new technology is needed and being actively deployed—that is, in emerging economies. Rich countries that have already met most of their energy needs will provide fewer opportunities than the late-industrializing countries for large-scale diffusion and improvement of new energy technologies, especially those providing large base-load generation and distribution. Improved coordination between rich countries, with established innovation capabilities, and later-industrializing nations, which are rapidly building out their energy systems, on improving clean energy technologies will both accelerate innovation and help to universalize equitable access to cleaner energy. Alleviating energy poverty therefore creates highly favorable conditions for the develop-

---

[20] Nasdaq, "U.S. National Average Gas Price," available at: http://www.nasdaq.com/markets/natural-gas.aspx?timeframe=10y

[21] Alex Trembath, Michael Shellenberger, Ted Nordhaus, and Max Luke, "Coal Killer: How Natural Gas is Fueling a Clean Energy Revolution," Breakthrough Institute (June 2013).

ment and diffusion of affordable low-carbon energy technologies.

We will discuss the opportunities for energy innovation in future chapters. In the meantime, China offers an obvious real-world example. It has demonstrated significantly lower costs for capturing carbon emissions from power plants than any other country.[22] It is a world leader in cost-effective hydroelectric power.[23] China has at least 20 nuclear reactors under construction (with 36 in full commercial operation), compared to four in the United States, and it is taking the lead in pushing forward advanced nuclear technologies like gas-cooled pebble-bed reactors and molten salt thorium-fed reactors.

Innovations that improve performance and cost through technology deployment (though not guaranteed[24]) are much more likely to accrue to China than to countries where energy demand is flat and energy systems are technologically, politically, and economically entrenched. And though the scale and intensity of effort in China is exceptional, other industrializing countries offer comparable opportunities. With sustained and coordinated financial, institutional, and technical support, similar development in support of improved energy access can happen in the least-developed countries, providing the foundation for a truly global energy innovation system.

---

[22] Robert C. Marley, "U.S.-China Clean Energy Research Center," presentation to the MIT Club of Washington (4 Feb. 2013), available at: http://www.us-china-cerc.org/pdfs/Marlay_US-China_CERC_MIT_4-FEB-2013_v10.pdf

[23] International Renewable Energy Agency, Hydropower: Renewable Energy Technologies: Cost Analysis Series, Vol. 1: Power Sector, Issue 3/5 (Bonn, Germany: IRENA Working Paper, June 2012).

[24] Arnulf Grübler, "The costs of the French nuclear scale-up: A case of negative learning by doing," *Energy Policy*, Vol. 38 (2010): pp. 5174-5188.

Training in China for the operation of the Westinghouse AP1000 nuclear reactor. *(Photo credit: NRC.)*

A final point on the global energy system: the variety of energy consumers and the services they require highlight how crucial it is for energy access and innovation to be context-appropriate. Energy needs are dependent on climate, geography, culture, economics, and a range of other factors. Urban and industrialized China, with its huge coal reserves, may profitably invest in carbon capture and storage technologies for its base-load coal power plants; the dispersed rural populations of Ethiopia may depend on extending grid infrastructure in order to exploit the country's abundant hydropower resources. And for remote communities around the world, distributed solar or wind generation and micro-grids may prove critical in providing near-term access to energy. The sheer diversity of contexts in which energy expansion will take place demands technological pluralism, meaning a commitment to moving innovation forward on nuclear,[25] fossil fuels, hydropower, solar, wind, transportation, infrastructure, and all other energy sources and services that can be made more affordable, cleaner, and socially acceptable.

[25] Joyashree Roy and Shyamasree Dasgupta, "The Economics of Nuclear Energy: Revisiting Resurrection," *Artha Beekshan*, Vol. 19, No. 1 (June 2010).

4

A HIGH-ENERGY PLANET

Energy modernization provides the foundation for future innovations and technology options that will lead to an increasingly clean global energy system. Furthermore, as societies increase their reliance on electricity and fuels with greater energy content and less carbon, delivered through efficient grid systems, there are substantial positive impacts on human well-being, economic productivity, and local environment—the pillars of sustainable development. Any morally acceptable and politically coherent path to reconcile energy access with successful climate action must be pursued not by minimizing energy consumption, but with the catalytic combination of truly equitable energy access and more energy innovation.

Because it generates greater human and capital resources with which to innovate, modern energy-supported development advances societies and their energy systems along trajectories distinct from, and ideally lower-carbon than, those traversed by the rich world.[1] This framework for development also enriches the potential for international, collaborative innovation efforts,

[1] Rob Byrne, Adrian Smith, Jim Watson, and David Ockwell, *Energy Pathways in Low-Carbon Development: From Technology Transfer to Socio-Technical Transformation* (Brighton, UK: STEPS Centre, STEPS Working Paper 46, 2011).

since the contexts for energy services, low-carbon innovation, and equitable access vary tremendously worldwide. The legitimate starting place for such collaboration is an explicit commitment to the kind of energy equity that enables an escape from subsistence living and fosters the capacity to prosper, adapt, and innovate.

The certain and irreversible global growth in energy consumption in turn provides the foundation for accelerating the technological, financial, and institutional innovations necessary to speed the transition to a low-carbon global energy system. Most energy consumption growth will come from later-industrializing countries as they fulfill their development ambitions. Ensuring that the least-developed countries, and poor and marginalized communities worldwide, benefit equitably from this expansion will remain a challenge. Yet with planning and imagination, escalating consumption is precisely—and counterintuitively—the process through which energy systems will develop along just and progressively lower-carbon pathways. By taking advantage of two global trends—the shift to an urbanized planet and the extraordinary growth of the energy sector in the developing world—we discover a pragmatic, plausible, and inclusive route to universalizing energy access and innovating toward low- or no-carbon energy futures.

Moving toward a high-energy planet is a moral imperative. By building on urbanization and energy-sector expansion in developing countries, we discover greater opportunities for achieving human development goals. Pursuing this agenda will not be easy or inexpensive, as there is no simple answer to difficult decisions about how to invest limited human, economic, and technical resources. But a high-energy planet is inherently enfranchising, empowering, and optimistic, and it works with rather than against the momentum of ongoing changes in an industrializing and urbanizing world. Our vision of a high-energy world can thus appeal to broad and diverse constituencies in ways that continue to escape the standard low-energy approach to climate policy.

By recasting engagement with the developing world within a high-energy framework, we create a foundation from which socioeconomic development and cleaner energy trajectories can be pursued. Groups and governments seeking to productively engage with the least-developed countries must address sector-wide energy problems through technical assistance, subsidy support, financing, and institutional capacity building, in order to help these countries take full advantage of their energy resources for the benefit of their citizens.[2] In rapidly growing economies like China, Brazil, and India, a focus on partnering with innovative energy institutions will be key to creating and massively diffusing the innovations that will decarbonize the global energy system and make energy available and affordable for all consumers. Such effective collaboration is the subject of the next section, on innovation for a high-energy planet.

It will take tremendous effort, capital, and political will to ensure that the ongoing expansion of the energy sector in developing nations provides all people with access to energy they can afford, and to support efforts that will make that energy progressively cleaner. But a high-energy framework aligns itself with the trends that are shaping the planet's future, improving the outcomes of these forces rather than futilely trying to stop or reverse them.

A framework for expanding energy access is useless if it neglects the rapidly urbanizing global population, the benefits of modern electricity grids and energy delivery systems, and the moral imperative of energy abundance and equity. Our vision of a high-energy planet is one in

---

[2] World Bank, *Addressing the Electricity Access Gap* (Washington, DC: World Bank Group, June 2010).

which human equity and well-being are top priorities, in which energy access is critically linked to governance and broader socioeconomic development, and in which the consumption and technology preferences of the rich world do not limit the ambitions and growth of developing countries.

Gary Dirks, Loren King, Frank Laird, Jason Lloyd, Jessica Lovering, Ted Nordhaus, Roger Pielke Jr., Mikael Román, Daniel Sarewitz, Michael Shellenberger, Kartikeya Singh, and Alex Trembath

Consortium for Science, Policy & Outcomes

DECEMBER 2014

BREAKTHROUGH INSTITUTE

5

INNOVATING TO ZERO

Energy consumption is essential to human development. Global energy use will thus increase significantly over the next century as poor nations achieve higher living standards. This is an overwhelmingly positive process in terms of life expectancy, health, and quality of life. Higher levels of energy consumption will also have significant environmental impacts. Some of the effects will be positive, as electricity and liquid fuels allow people to move away from wood and dung as primary fuels, which contribute to respiratory disease and deforestation. At the same time, rising fossil fuel consumption results in high levels of air pollution in rapidly growing cities and contributes to global warming, with potentially large economic and environmental costs.[1]

---

[1] Intergovernmental Panel on Climate Change (IPCC), "Summary for Policymakers," in *Climate Change 2014: Impacts, Adaptation, and Vulnerability. Part A: Global and Sectoral Aspects. Contribution of Working Group II to the Fifth Assessment Report of the Intergovernmental Panel on Climate Change*, C. B. Field, V. R. Barros, D. J. Dokken, K. J. Mach, M. D. Mastrandrea, T. E. Bilir, M. Chatterjee, K. L. Ebi, Y. O. Estrada, R. C. Genova, B. Girma, E. S. Kissel, A. N. Levy, S. MacCracken, P. R. Mastrandrea, and L. L. White, eds. (New York, NY: Cambridge University Press, 2014), pp. 1-32.

Past energy transitions show a trend toward cheaper, cleaner, more abundant, and more reliable new fuels, as well as the replacement of old energy technologies with new ones. For more than 200 years, public and private actors have worked to move nations up the "energy ladder," from wood, dung, and charcoal to diversified modern systems consisting of fossil fuels like coal, oil, and natural gas and low-carbon technologies like hydroelectricity, nuclear, and renewables.[2] This is a long-term trend toward less pollution and fewer carbon emissions. To be sure, every nation's geography, energy reserves, and technical capacities differ, and so each national energy modernization process is unique. But the collective desire for cheaper, cleaner, and more reliable energy is behind this emergent global phenomenon of decarbonization.

Given the importance of climbing the energy ladder for human development, continuous technological innovation of energy systems has been a priority for prosperous nations since the Industrial Revolution. Rich nations, in turn, have understood their global role in linking energy to human development as one of "transferring" technologies to poor countries.[3] This model has been directly applied in international efforts to address global warming, for example, through the United Nations Framework Convention on Climate Change.[4]

But in the last decade, the center of gravity in energy innovation has shifted decisively to rapidly industrializ-

---

[2] Roger Fouquet and Peter J. G. Pearson, "Past and prospective energy transitions: insights from history," *Energy Policy*, Vol. 50 (2012): pp. 1-7.

[3] Kelly Sims Gallagher, *The Globalization of Clean Energy Technology: Lessons from China* (Cambridge, MA: MIT Press, 2014).

[4] Stephen Seres, Erik Haites, and Kevin Murphy, "The Contribution of the CDM under the Kyoto Protocol to Technology Transfer," United Nations Framework Convention on Climate Change (2010).

ing countries. These nations dominate the manufacturing of solar, wind, biofuel, and other technologies, and are rapidly deploying and innovating on nuclear power, hydroelectricity, and natural gas.[5]

This should be unsurprising: innovation tends to occur where demand for new technologies is growing fastest, and energy is no exception.[6] As most of the new energy infrastructure over the coming decades will be built in industrializing countries,[7] it is there that we should expect to see—and should work hardest to accelerate—energy innovation. This innovation will produce global economic and environmental benefits, as cheaper energy technologies literally fuel productivity gains across all sectors of society. Thus we argue that clean energy innovation is a global public good to be pursued collaboratively by nations seeking to advance their economic, social, and environmental well-being.

In contrast to our view that clean energy innovation requires international collaboration, a number of analysts and policymakers over the past decade have framed energy innovation as a "clean tech race," a zero-sum game played by nations competing to dominate low-carbon energy industries for domestic economic advantage.[8]

---

[5] Rob Atkinson, Michael Shellenberger, Ted Nodhaus, Devon Swezey, Teryn Norris, Jesse Jenkins, Leigh Ewbank, Johanna Peace, and Yael Borofsky, *Rising Tigers, Sleeping Giant* (Oakland, CA and Washington, DC: Breakthrough Institute and Information Technology & Innovation Foundation, 2009).

[6] D. Mowery and N. Rosenberg, "The influence of market demand upon innovation: a critical review of some recent empirical studies," *Research Policy*, Vol. 8 (1979): pp. 102–153.

[7] U.S. Energy Information Administration, *International Energy Outlook 2013* (Washington, DC: DOE/EIA-0484, July 2013).

[8] For example, see: Pew Charitable Trusts, "Who's Winning the Clean Tech Race?" (April 2014).

This view was reinforced in the early 2010s by trade disputes over solar panel manufacturing. Efforts by China, the United States, and the European Union to accelerate the deployment of solar power helped drive down costs, but also sparked an international trade war, as manufacturers in rich countries could not compete with cheap Chinese panels.[9] Such competitive framing is ultimately self-defeating. The economic benefits that flow to individual countries by being competitive in manufacturing advanced energy technologies are small compared to the overall public benefits (including economic) of energy that is both cheap and clean.

As such, the crucial yet complex role of energy innovation in global development needs to be reconceived from the bottom up. A new and empowering understanding starts with the recognition that opportunities for energy innovation and decarbonization on our high-energy planet are concentrated in rapidly industrializing economies. For wealthy countries to contribute decisively, they will need to play a different role than either technology provider or economic competitor. We argue that rising energy consumption is an opportunity to advance both human development and environmental protection through pragmatic policies. Chief among them is technological innovation to make energy cheaper, cleaner, more reliable, and more abundant.

---

[9] Jeffrey Ball, "The Battle in Our Trade War with China," *The New Republic* (21 Jan. 2014).

6

THE RISE OF THE REST

Transformative technologies are rarely invented in the research laboratory and unveiled to a grateful world. Rather, new materials, processes, and physical phenomena are discovered both in and outside the laboratory. They are applied in new contexts, tinkered and combined with other technologies, sometimes in research laboratories but mostly in the real world—be it a factory floor, battlefield, hospital operating room, or farm. The study of innovation over the past several decades, across multiple contexts, economic sectors, and stages of technology development and use, has consistently concluded that processes of invention and innovation are not linear. Invention, innovation, diffusion, and use feed back into and depend on one another in complex, indirect, and unpredictable ways.[1]

These observations are illustrated by the rise of the Internet and the World Wide Web, two innovations that have revolutionized our world. These drivers of social and economic change were not designed from scratch or even imagined far in advance. They emerged over many decades from advances in information and communications technologies, in network theory and other fundamental

---

[1] R. R. Nelson and S. G. Winter, "Evolutionary Theorizing in Economics," *The Journal of Economic Perspectives*, Vol. 16, No. 2 (2002): pp. 23–46.

sciences. Above all, today's Internet and the web are the result of the demands, ingenuity, and experience of users, from scientists in academic laboratories, to entrepreneurial individuals and firms looking for new products and markets, to government agencies trying to better deliver services and information.[2]

The implications of these dynamics are significant for global energy innovation efforts. If energy technology deployment over the coming decades is overwhelmingly concentrated in developing economies, then that is where most energy technology innovation will likely occur. Innovation activities that are divorced from or not well integrated with the sites of deployment and use are likely to fail. Furthermore, because a nation's capacity to innovate and deliver abundant, cheap energy across its economy are inextricable from broader processes of socioeconomic advancement, energy innovation efforts must be grounded in and contribute to ambitious development agendas.

These dynamics challenge the long-standing framework for global energy innovation. Dating back to the famous UN-commissioned Brundtland Report,[3] which in 1987 articulated a vision for pursuing global sustainability, this framework helped set the agenda for international energy and environmental initiatives. It imagined that poor countries, through the transfer of low-carbon energy technologies from rich nations, could develop their energy systems along trajectories that are radically different than those of earlier-industrializing societies.

---

[2] Jesse Jenkins, Devon Swezey, and Yael Borofsky, "Where Good Technologies Come From: Case Studies in American Innovation," Breakthrough Institute (5 Dec. 2010).

[3] Gro Harlem Brundtland, *World Commission on Environment and Development: Our Common Future* (Oxford, UK: Oxford University Press, 1987).

The Brundtland Report was a product of energy and development thinking dominant among well-meaning Westerners in the 1960s and 1970s. European and U.S. environmental and development critics, living in the wealthiest and most secure political economies in history, disavowed the modernization pathways their countries had followed. To avoid global environmental, economic, and demographic catastrophe, these critics claimed, poor countries could not follow that example.[4] Influenced strongly by E. F. Schumacher's "appropriate technology" prescriptions,[5] Amory Lovins's warnings against energy consumption and centralized energy systems,[6] and the Club of Rome's dire projections of global resource shortages,[7] a new framework emerged: the soft-energy paradigm.

This framework is predicated on two core assumptions. The first is that "a low-energy path is the best way towards a sustainable future," as the Brundtland Report insists. The second is that existing renewable energy technologies will replace most fossil fuel use, obviating the need for substantial innovation in clean energy systems.[8]

---

[4] Martin W. Lewis, *Green Delusions: An Environmentalist Critique of Radical Environmentalism* (Durham, NC: Duke University Press, 1992).

[5] Ernst Freidrich Schumacher, *Small is Beautiful: Economics as if People Mattered* (London, UK: Blond & Briggs, 1973).

[6] Amory B. Lovins, "Energy Strategy: The Road not Taken," *Foreign Affairs*, Vol. 55 (1976): pp. 65-96.

[7] Donella H. Meadows, Jorgen Randers, Dennis L. Meadows, and William W. Behrens, *The Limits to Growth: A Report for the Club of Rome's Project on the Predicament of Mankind* (New York, NY: Universe Books, 1972).

[8] Megan Nicholson and Matthew Stepp, *Challenging the Clean Energy Deployment Consensus* (Washington, DC: Information Technology & Innovation Foundation, 2013).

The Brundtland framework provided the normative principle for the United Nations and its Framework Convention on Climate Change (UNFCCC), the main instrument by which the international community endeavors to mitigate the climate impact of human activities.[9] It is also the paradigm for low-carbon development initiatives like the Global Environment Facility (GEF), Climate Investment Funds (CIF), and Clean Development Mechanism (CDM).[10] The UNFCCC expresses the Brundtland Report's conviction that poor countries could assume novel development pathways through minimized energy consumption and renewable energy deployment, especially through provisions that allow rich countries to meet their emissions reduction commitments most cost effectively by supporting low-carbon projects in developing countries.[11]

Unfortunately, approaching energy system development in poor countries with a single-minded focus on non-emitting renewables—energy technologies with significant limitations for meeting the needs of energy-starved, rapidly urbanizing developing countries—undermines the creation of a robust, diversified energy infrastructure. Off-grid renewables can in some cases provide limited energy access more quickly or cheaply than

---

[9] John Drexhage and Deborah Murphy, "Sustainable Development: From Brundtland to Rio 2012," Background Paper for the High Level Panel on Global Sustainability (New York, NY: United Nations, Sept. 2010).

[10] Rob Byrne, Adrian Smith, Jim Watson, and David Ockwell, *Energy Pathways in Low-Carbon Development: From Technology Transfer to Socio-Technical Transformation* (Brighton, UK: STEPS Centre, STEPS Working Paper 46, 2011).

[11] Dieter Helm, *The Carbon Crunch: How We're Getting Climate Change Wrong – and How to Fix It* (New Haven, CT: Yale University Press, 2012), p. 164.

conventional baseload power and grid expansion.[12] But the priorities of energy system expansion efforts in the developing world, and the donor countries and organizations that work there, must be consistent with broader development objectives that include agricultural modernization, the creation of domestic industrial capacity, and meeting the needs of rapidly growing cities. Powering the development of contemporary urban, agricultural, and industrial infrastructures requires large quantities of cheap, baseload power and liquid fuels, as discussed in earlier chapters.

An analysis from the Center for Global Development compares access rates in sub-Saharan Africa between two hypothetical $10 billion energy project investment portfolios. One comprises only renewables and the other contains only natural gas projects. The gulf in access rates is enormous: "A natural gas-only portfolio could provide electricity access to 90 million people versus 20 to 27 million people with a renewables-only portfolio." A project investment portfolio of two-thirds natural gas projects and one-third renewables would support energy access for 70 million people, or at least 40 million more than renewables alone.[13]

As this analysis demonstrates, simply transferring existing renewable technologies to developing countries cannot provide the energy necessary for development. Nor can this mechanism catalyze the economic activities necessary to spur indigenous capacities for technological

---

[12] Benjamin Leo, Vijaya Ramachandran, and Robert Morello, "Shedding New Light on the Off-Grid Debate in Power Africa Countries," Center for Global Development (14 Oct. 2014).

[13] Todd Moss and Benjamin Leo, "Maximizing Access to Energy: Estimates of Access and Generation for the Overseas Private Investment Corporation's Portfolio," Center for Global Development (Jan. 2014).

innovation. Perhaps, then, rich countries will develop the low-carbon hardware necessary both to leapfrog fossil energy use and to power high-energy economies, and transfer this next generation of innovative energy technology to the developing world?

The reality is that wealthy economies are unlikely to offer either the motivation or context in which rapid clean energy innovation might occur. In developed countries, energy demand projections are flat or decreasing. With energy infrastructure and transitions lasting several decades at least, it makes little economic sense for developed nations to make large investments in clean energy innovation. Power plants in the United States have a replacement cost of $1.5 trillion.[14] Sunk costs are a tremendous incentive against disruptive innovation.[15] Of the wealthy nations, only Germany and Denmark are making a comprehensive effort to transform their energy systems to low-carbon ones, and the outcomes of those experiment are both highly uncertain and far in the future.[16]

National interest has often played a key role in driving innovation. The United States' development of light-water nuclear reactors was borne out of defense concerns, with the design originally created for military submarines.[17] The original funding for shale gas exploration—investment that kick-started a decades-long process that

---

[14] Vaclav Smil, "Moore's Curse and the Great Energy Delusion," *The American* (19 Nov. 2008).

[15] Clayton M. Christensen, *The Innovator's Dilemma: When New Technologies Cause Great Firms to Fail* (Cambridge, MA: Harvard Business Review Press, 1997).

[16] Will Boisvert, "Green Energy Bust in Germany," *Dissent Magazine* (Summer 2013); Christine Sturm, "Inside the *Energiewende*: Policy and Complexity in the German Utility Industry," *Issues in Science and Technology*, Vol. 33, No. 2 (Winter 2017).

[17] Jenkins, et al., "Where Good Technologies Come From."

ultimately led to fracking—was justified by U.S. concern with its dependency on foreign oil.[18] Energy independence was also a reason for France and Sweden's rapid transitions to nuclear.[19]

Figure 2: World Primary Energy Consumption, 1990-2040

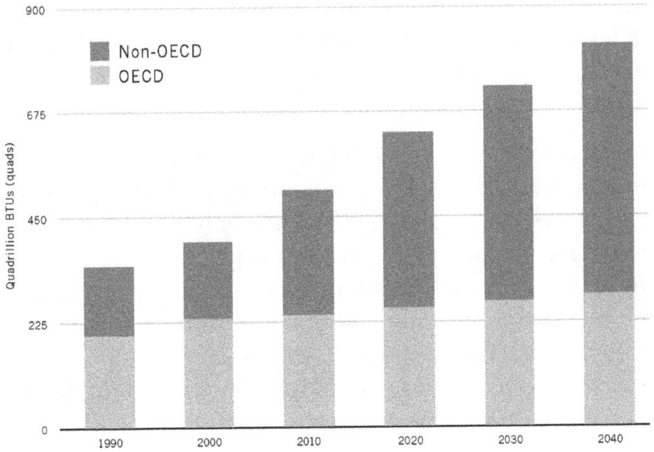

Source: U.S. Energy Information Administration, *International Energy Outlook 2013.*

At times, energy innovation can be a means to gain a comparative advantage in international trade, as has been suggested of China's recent push into solar photovoltaic (PV) manufacturing. But at its core, countries are driven to innovate in the energy domain because cheap, reliable (which often means domestically produced), and abundant energy is essential to economic growth and national

---

18 Alex Trembath, Jesse Jenkins, Ted Nordhaus, and Michael Shellenberger, "Where the Shale Gas Revolution Came From," Breakthrough Institute (2012).

19 Jesse Jenkins, "Historic Paths to Decarbonization," Breakthrough Institute (3 April 2012).

prosperity. For most developed countries, cheap and abundant energy already exists.

By contrast, rapidly industrializing countries are power hungry. As illustrated in Figure 2, nearly all of the growth in energy markets and the majority of new energy technologies deployed in the coming decades are projected to occur in the developing world. This is a direct result of building out energy systems to support development ambitions and provide citizens with access to the energy they need to prosper.

Societies that have successfully accelerated their development have done so by expanding modern diversified energy systems, building the knowledge and experience necessary for improved performance and continual learning in the process.[20] The pattern of gradually strengthening innovation capacity, specific to historical and national contexts, has been central to the modernization of every industrialized and industrializing nation, from England to the United States to South Korea to Brazil.[21]

All of the components of a country's energy system—power plants, pipelines, electricity grids, and so on—are tightly interdependent. In other words, energy systems are an example of "technological lock-in," where complementarities between individual technologies and infrastructure are very strong.[22] This locked-in aspect of a

---

[20] Martin Junginger, Wilfried van Sark, and André Faaij, eds., *Technological Learning in the Energy Sector: Lessons for Policy, Industry and Science* (Northampton, MA: Edward Elgar, 2010).

[21] Nathan Rosenberg, "The International Transfer of Technology: Implications for the Industrialized Countries," Ch. 11 in *Inside the Black Box: Technology and Economics* (Cambridge, UK: Cambridge University Press, 1982).

[22] W. Brian Arthur, "Competing technologies, increasing returns, and lock-in by historical events," *The Economic Journal*, Vol. 239, No. 7 (July 2012).

nation's energy system means that technological innovations that fit relatively seamlessly into the existing regime are adopted far more quickly than those that do not. This is why, as we discuss in more detail below, the fracking revolution occurred in the United States and accounted for speedy reductions in carbon emissions, in contrast to the much slower diffusion of renewables. Fracking was made possible by incremental improvements of existing hydrocarbon extraction technologies, and the resulting natural gas could be incorporated into existing energy infrastructure.[28] Such "path dependencies" are characteristic of modern, locked-in energy systems.

Figure 3: Carbon-Free Energy as Portion of Added Energy Consumption, 1966-2012

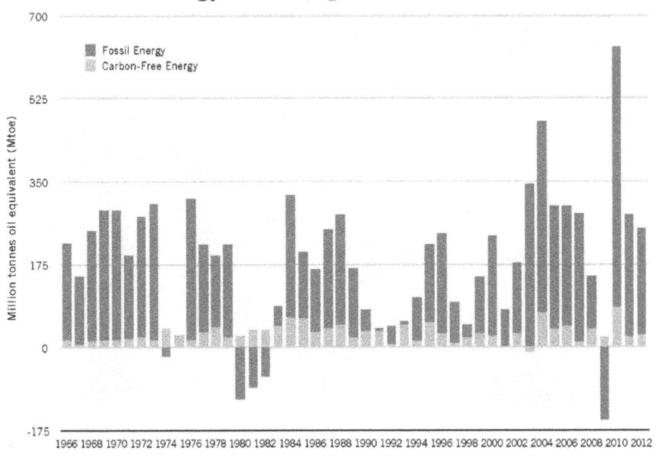

Source: BP, "Statistical Review of World Energy, 2012."

If rich countries are constrained by technological lock-in and path dependencies, in developing countries the relative lack of preexisting infrastructure means energy innovation can explore new and diverse technology pathways as they build out their energy systems to meet their economic and social development needs. This presents both an opportunity and a challenge: on the one hand, developing countries are less invested in the prevailing fossil fuel regime. On the other, developing countries will continue to exploit fossil fuels as the most efficient path to modernization.

No country has succeeded in achieving significant human development or economic growth without a leading role for fossil fuels, along with other modern technologies like large hydroelectric power.[23] To date, there are no countries even attempting to pursue a development path similar to that encapsulated by the Brundtland's low-energy framework. The world's growth in fossil fuels consumption is still far outpacing that of clean energy (see Figures 3 and 4).

Figure 4: Projected Global Primary Consumption, 2010-2050

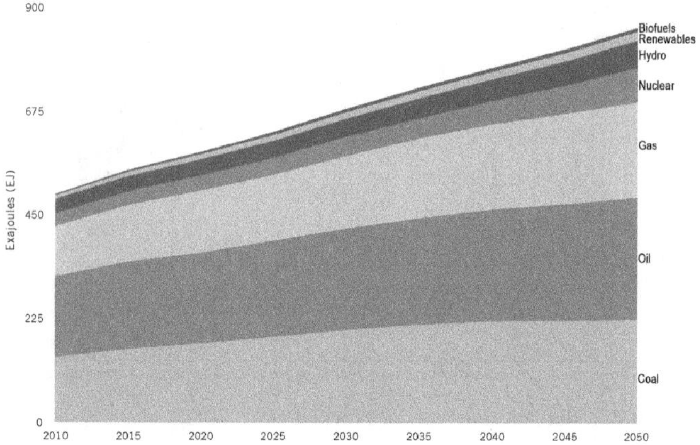

Source: MIT Joint Program on the Science and Policy of Global Change, "MIT Energy and Climate Outlook 2014."

---

[23] Arthur Van Benthem, "Has Energy Leapfrogging Occurred on a Large Scale?" United States Association for Energy Economics and the International Association for Energy Economics, Working Paper (2010).

Yet the sheer scale of providing the energy necessary to power economic and social growth has compelled developing countries to invest in a wide range of technologies. Whether it is experiments with renewables and storage in the United Arab Emirates' Masdar City,[24] grid expansion in Brazil,[25] or underground coal gasification in South Africa,[26] industrializing countries are not restricting themselves to conventional fossil fuels. Indeed, developing countries may transition to advanced energy systems faster, with a greater variety of energy sources, and more efficiently than was the case in the United States.[27]

China, in particular, is investing heavily in clean energy, partly as a means to gain a competitive advantage but mostly to pursue an "all-of-the-above" strategy and deal with mounting pollution problems in its cities. The country is pioneering fourth generation nuclear reactors, such as sodium-cooled fast reactors, high-temperature gas reactors, and salt-cooled reactors.[28] Combined, the emerging economies of Brazil, Russia, India, Mexico, China, and South Africa provide as much public funding on energy research, development, and deployment as do all 29

---

[24] Jared Anderson, "Masdar City: New Urban Energy Future and Climate Change Solution?" *Breaking Energy* (20 Mar. 2013).

[25] John Lyons, and Paulo Trevisani, "Brazil, China Sign Power-Grid, Plane Deals," *Wall Street Journal* (17 July 2014).

[26] James Burgess, "South Africa to Use Large-Scale Underground Coal Gasification," Oilprice.com (28 April 2013).

[27] Arnulf Grübler, "Energy transitions research: Insights and cautionary tales," *Energy Policy*, Vol. 50 (2012): pp. 8-16; Peter J. Marcotullio and Niels B. Schulz, "Comparison of Energy Transitions in the United States and Developing and Industrializing Economies," *World Development*, Vol. 35, No. 10 (2007): pp. 1650-1683.

[28] Ted Nordhaus, Jessica Lovering, and Michael Shellenberger. "How to Make Nuclear Cheap: Safety, Readiness, Modularity, and Efficiency," Breakthrough Institute (July 2013).

wealthy member countries of the International Energy Agency.[29]

Of course large-scale investments in clean energy are not occurring evenly or equally across the developing world. Clean energy innovation requires a robust industrial base, with easy access to both suppliers and consumer markets. In most developing countries, the process of industrialization is still in its infancy and research and manufacturing capacities remain modest—weaknesses that will be ameliorated as these countries work to expand their energy systems. They are doing this in part with help from affluent donor nations, but mostly (and most pragmatically) with the assistance of rapidly developing countries, most notably China.

Our focus is thus squarely on rapidly industrializing countries. Substantial research, commercial, trade, and investment potentials already exists in these countries. Coupled with growing demand for essentially everything—but especially energy—it is in industrializing countries that policymakers and donors should target interventions aimed at advancing and accelerating clean energy innovation.

---

[29] Kelly Sims Gallagher, Arnulf Grübler, Laura Kuhl, Gregory Nemet, and Charlie Wilson, "The Energy Technology Innovation System," *Annual Review of Environment and Resources*, Vol. 37 (2012), pp. 137-162.

7

GLOBAL INNOVATION ECOSYSTEMS

In this chapter, we evaluate innovation progress on four energy technologies with the potential to provide cheap, clean, and reliable baseload power through rapid deployment in industrializing economies. We focus on these four not to suggest that they should be the only energy technologies pursued by international efforts, but rather to illustrate the distinct challenges facing different technologies, including their innovation and diffusion in different national contexts.

Shale Gas

The recent boom in natural gas production in the United States, brought about through technical innovations in the recovery of natural gas from previously inaccessible shale rock formations and land-use policies that favor private development, has helped lower electricity costs and benefitted the petrochemical and manufacturing industries.[1] Even more significantly, it has contributed to a drop in U.S. carbon dioxide emissions to their lowest levels in two

---

[1] Richard Nemec, "U.S. Reindustrialization an Offshoot of Natural Gas Boom," *Pipeline & Gas Journal*, Vol. 239, No. 7 (July 2012).

decades,[2] as inexpensive natural gas accelerates the closure of aging coal plants around the country.

Though hydraulic fracturing's diffusion across the United States since 2005 was rapid,[3] the actual innovation process occurred over decades. The technique of fracturing rock to recover fuels was invented in the late 1940s, but it required many additional innovations—the result of public-private partnerships and federal investments at many points in the process—to develop a method of fracking that was economically viable.[28] The version of fracking that came to dominate was the one that took advantage of resources available to U.S. companies, particularly the abundant water supplies that made feasible injecting millions of gallons of water into underground rock formations.[43] Fracking's economic success also depended on external factors such as continuous improvements to the country's energy infrastructure, especially its natural gas pipelines.

The possibility of cheaper and cleaner energy from shale gas has prompted interest from governments around the world. If it can achieve the necessary innovations for tapping perhaps the largest shale gas reserves on the planet, China may be able to reduce its dependence on coal and shift to a lower-carbon economy.[4] European

---

[2] Russell Gold, "Rise in U.S. Gas Production Fuels Unexpected Plunge in Emissions," *Wall Street Journal* (18 Apr. 2013).

[3] Z. Wang and A. Krupnick, "A Retrospective Review of Shale Gas Development in the United States: What Led to the Boom?" *SSRN Electronic Journal* (2013).

[4] Lei Tian, Zhongmin Wang, Alan Krupnick, and Xiaoli Liu, *Stimulating Shale Gas Development in China: A Comparison with the U.S. Experience* (Washington, DC: Resources for the Future Discussion Paper 14-18, July 2014).

countries such as the United Kingdom are also exploring the possibility of exploiting shale gas.[5]

However, caution is warranted. The large deployment of fracking technology faces significant hurdles outside of the U.S. context. China's nascent industry is plagued by technical bottlenecks, lack of adequate water supply, and poor infrastructure.[6] Drilling an exploratory shale gas well in China still costs much more than it does in the United States.[7] In Europe, the challenges are more likely to be political and legal.[8] Unlike in the United States, European landowners do not automatically own the rights to extract the resources from the ground beneath their property, making the building of new extraction plants fraught with political difficulties.[9]

From this example, three lessons are clear. First, incremental innovation within an existing and powerful segment of the energy sector has lowered American carbon emissions and reaped substantial benefits to the economy. The shale gas revolution has reduced U.S. power sector emissions on the order of 150 to 200 megatons

---

[5] Paula Dittrick, "UK government lifts hydraulic fracturing ban," *Oil and Gas Journal* (13 Dec. 2012).

[6] Desheng Hu and Shengqing Xu, "Opportunity, challenges and policy choices for China on the development of shale gas," *Energy Policy*, Vol. 60 (2013): pp. 21-26.

[7] Alan Krupnick, presentation at "The Future of Fuel: Toward the Next Decade of U.S. Energy Policy," RFF First Wednesday Seminar, Resources for the Future (28 Nov. 2012).

[8] Dieter Helm, *The Carbon Crunch: How We're Getting Climate Change Wrong – and How to Fix It* (New Haven, CT: Yale University Press, 2012), p. 164.

[9] Russell Gold, *The Boom: How Fracking Ignited the American Energy Revolution and Changed the World* (New York, NY: Simon & Schuster, 2014).

1970s-2000

1. The Eastern Gas Shales Project was a federally funded shale fracking research and development and demonstration project that constructed test wells in major shale plays in the eastern **US**.

2. The Western Gas Shales Project was a federally funded unconventional oil and gas research, development, and demonstration project in southwestern states including **Colorado** and **New Mexico**.

3. Sandia and Los Alamos created the Natural Gas & Oil Technology Partnership, which eventually included work from several other **US** Department of Energy labs. Sandia was instrumental in developing and sharing critical microseismic imaging technology with Mitchell Energy.

4. Los Alamos National Labs in **New Mexico** performed significant research, development, and testing of early microseismic imaging technologies for application in both oil and gas exploration and initially for the Lab's Geothermal Program.

5. The **US** Department of Energy funded and executed much of the major groundbreaking unconventional gas innovation from the 1970s through the 2000s. Congress also created the Section 29 tax credit for unconventional gas that lasted from 1980 to 2002.

6. The Barnett Shale in northeast **Texas** was the site of Mitchell Energy's major technological breakthroughs in shale gas fracking in the 1980s and 1990s.

2000-today

7 Chevon, Cuadrilla (UK), and other multinational O&G firms have invested with Polish companies in exploration of Polish shale energy resources. State-controlled energy companies PGNiG and PKN Orlen are leading shale gas exploration in **Poland**.

8 Total, Shell, and Exxon Mobil have all invested in **Argentina's** Vaca Muerta shale formation. Chevron and YPF, Argentina's state oil company, have a joint venture to develop shale resources.

9 In 2012 **South Africa** lifted a national fracking moratorium, opening investment and exploration opportunity for international oil companies, including Royal Dutch Shell.

10 France-based Total became the first major multinational to enter into shale gas exploration in the **UK**. Cuadrilla is drilling dozens of exploratory wells there.

11 **China's** Sinopec bought $2.5 billion stake in Devon Energy in 2012. Other Chinese O&G firms have bought stakes in US oil production projects developed by Chesapeake.

12 Shale gas was the top priority at the **US-India** Energy Partnership Summit, attended by new Indian PM Narendra Modi, in September 2014.

13 **Ukraine** has signed agreements with Royal Dutch Shell and Chevron for shale gas exploration.

annually over the past decade, and cheaper energy costs have provided a $100 billion-per-year boost to the United States economy.[10] Second, the diffusion of energy technologies beyond the techno-economic system from which they emerge is rife with challenges. Third, and precisely because this process is so hard, the transfer of expertise and technical knowledge (rather than merely dropping in hardware) is critical to accelerating diffusion.

Countries have tried to do this by attracting the expertise of American firms. Mexico, for example, has opened up its oil and gas sector to foreign investment[11] in order to acquire the horizontal drilling and hydraulic fracturing techniques that can help it access one of the world's largest reserves of shale gas and tight oil.[12] And a Chinese energy company, Sinopec Group, paid Devon Energy (which had previously acquired Mitchell Energy, the firm that co-created the shale gas revolution with the U.S. government) billions of dollars to work with it on fuel extraction projects, in the hope of gaining access to the American firm's expertise.[13] Other countries are enthusiastically exploring the possibility of shale gas production, including Argentina, South Africa, and Poland.[14]

---

[10] Alex Trembath, "U.S. Coal Exports Do Not Offset Massive Emissions Reductions from Natural Gas," Breakthrough Institute (19 Aug. 2014).

[11] Jude Webber, "Mexico courts foreign investment with energy reform," *Financial Times* (12 Dec. 2013).

[12] Michael Moran, "Peso Me Mucho: Why Mexico is the new darling of emerging markets," *Foreign Policy* (29 Oct. 2013).

[13] Jim Polson and Benjamin Haas, "Sinopec Group to Buy Stakes in Devon Energy Oil Projects," *Bloomberg* (3 Jan. 2012).

[14] Roberto F.Aguilera and Marian Radetzki, "Shale gas and oil: fundamentally changing global energy markets," *Oil & Gas Journal*, Vol. 111, No. 12 (2013).

Nuclear

Nuclear power is energy dense, provides reliable base-load power, and offers a range of highly advantageous end uses, such as the ability to generate large quantities of process heat for water desalination and other industrial purposes. But rising capital costs and systemic barriers to nuclear innovation over the past four decades have limited its ability to make a significant dent in fossil fuels' dominance.

Most of the growth in commercial nuclear power over the coming decades will occur in rapidly industrializing countries like China and South Korea, and Middle Eastern countries like Saudi Arabia and the United Arab Emirates. Indeed, 20 of the 58 reactors currently under construction in the world are being built in China.[15] By contrast, the dominant rich-world markets for nuclear power—including the United States, France, Sweden, and Japan—have either dramatically slowed their nuclear build-out or pursued a path of accelerated decommissioning, as in the case of Germany. And nuclear is unlikely to be an option in poor nations lacking strong scientific, technical, and regulatory establishments.

In the 1960s, conventionally constructed thermal reactors became "locked-in" as the dominant technology at the expense of other designs, including thorium-fueled, pebble-bed, gas-cooled, and fast reactors.[16] Five decades later, nuclear innovation is occurring with both conventional light-water reactors and next-generation reactors using new coolant and fuel designs. For instance, the Chinese Academy of Sciences is currently building on research

---

[15] Mycle Schneider and Antony Froggatt, *The World Nuclear Industry Status Report 2016* (Jul. 2016).

[16] Per Peterson, Micahel Laufer, and Edward Blandford, "Nuclear Freeze: Why Nuclear Power Stalled," *Foreign Affairs* (May/June 2014).

into a molten salt reactor (MSR), initiated and later discarded by the United States' Oak Ridge National Laboratory in the 1960s, with the aim of constructing a thorium-breeding MSR prototype in Shanghai by 2020. The U.S. Department of Energy is collaborating on the project, which reportedly has a start-up budget of $350 million.[17] Bill Gates has been in talks with the China National Nuclear Corporation about developing his idea for a traveling-wave reactor.[18]

Rapidly developing countries are leading the way on advanced reactor designs across the board. Russia has been operating sodium-cooled fast reactors since the 1980s, started construction on an 800MW commercial design,[19] plans to construct the same reactor in China,[20] and is beginning work on two different lead-cooled fast reactor demonstrations. India and China are also operating their own experimental fast reactors and planning for larger demonstrations. China has begun construction on a 210MW high-temperature gas reactor.[21]

The United States government could do more to facilitate international cooperation, governance and safety, and

---

[17] World Nuclear Association, "China's Nuclear Fuel Cycle" (20 Nov. 2013); Richard Martin, "China Details Next-Gen Nuclear Reactor Program," *Technology Review* (16 Oct. 2015).

[18] A traveling-wave reactor can breed its own fuel from fissile material like spent fuel from other reactors or natural uranium. See: "Bill Gates Discussing New Nuclear Reactor with China," Associated Press (7 Dec. 2011).

[19] "Fast reactor starts nuclear clean energy era in Russia," *RT* (27 June 2014).

[20] "China signs up Russian fast reactors," *World Nuclear News*, (15 Oct. 2009).

[21] International Atomic Energy Agency, "Support for Innovative Fast Reactor Technology Development and Deployment," IAEA (July 2014).

knowledge spillover. One example is the commercialization of fuel reprocessing. The United States and South Korea have cooperated on the development of civilian nuclear resources since the 1950s.[22] Nuclear currently provides around 40 percent of South Korea's electricity needs, and the country has recently become an exporter of nuclear technology.[23] However, proliferation concerns have made the United States reluctant to share research on reprocessing spent fuels, which has hindered South Korea's efforts to deal with waste disposal.[24] American leadership in reforming international governance regimes and coordinating other areas of research and demonstration would likely yield benefits not just to the countries actively involved in such projects, but also to future consumers of advanced nuclear technologies—the "late adopters" who can capitalize on early collaborative enterprises.

---

[22] Mark Holt, "U.S. and South Korean Cooperation in the World Nuclear Energy Market: Major Policy Considerations," Congressional Research Service (25 June 2013).

[23] Margaret Coker, "Korean Team to Build U.A.E. Nuclear Plants," *Wall Street Journal* (28 Dec. 2009).

[24] Sebastian Sarmiento-Saher, "South Korea's Nuclear Blues," *The Diplomat* (19 Jun. 2013).

GENIII+

1. To build its first nuclear power plant, the **United Arab Emirates** contracted with Korea Electric Power Co. The contract (signed in 2009), which includes equipment from Westinghouse, marks the first international deal for **South Korea's** fledgling nuclear export industry. The first of these reactors will come online in 2017.

2. Westinghouse's AP1000 dominates the international GenIII+ market, with dozens of projects underway in the **United States, China,** and **Bulgaria**. Westinghouse has 4 AP1000s under construction along the coast of **China**, and the multinational corporation is close to signing a deal for up to 26 additional nuclear plants to be built throughout **China's** interior.

3. Paris-based Areva and Berlin-based Siemens collaborated to design the European Pressurized Water Reactor (EPR) reactor, a competitor of the AP1000, which is under construction in **Finland**, **France**, and **China**. Another two EPRs are planned for the **UK** with an expected commercial start date in the early 2020s.

4. **Russia's** VVER reactor series has long operated in **Armenia, Bulgaria, China, Czech Republic, Finland, Hungary, India, Iran, Slovakia, Ukraine, and Russia.** Their new GenIII+ design, the VVER-1200, is targeted to countries hoping to build their first nuclear power plants. **Russia** offers the lowest cost GenIII+ design, while also helping with financing, liability, and managing the fuel cycle, which companies like Westinghouse and Areva may not offer.

GENIV

5. Rosatom's BN-800 reactor (a sodium-cooled fast neutron reactor) began producing electricity in **Russia** in October 2014, and the state-owned company is also working with the China Nuclear Energy Industry Corporation on a project of the same design in **Sanming City, China** making it the first fast reactor built for commercial export.

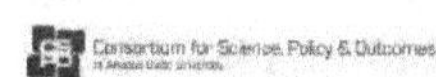

Nuclear Energy Innovation Collaboration

The United States, China, and other countries are working together to build the next generation of safer, cleaner, and cheaper nuclear power plants.

6 **China** is constructing a two-unit modular demonstration of a high-temperature gas-cooled reactor that will use a pebble bed fuel, the design of which was based on the German High-Temperature (HTR)-MODUL, which was never built in **Germany** due to antinuclear sentiment in the wake of Chernobyl. These reactors are scheduled to come online in 2017.

7 Through a knowledge-sharing agreement that began in 2010, the Chinese Academy of Sciences enlisted the US Oak Ridge National Laboratory for two thorium molten salt reactor experiments based on research pioneered by the **United States** in the 1960s and 1970s. **China's** 2 MW solid-fueled molten salt reactor should begin operation in 2017, while the liquid-fueled experiment won't start until 2020 at the earliest.

FUSION

8 The International Thermonuclear Experimental Reactor (ITER) in **France** is one of the largest collaborative energy projects ever built, involving more than **30 countries** investing more than $50 billion. The first plasma is expected to be created in 2020.

FUEL

9 Thor Energy, co-owned by the **South African** company Steenkampskraal Thorium, has started testing its own solid thorium MOX fuel rods in **Norway**, which are designed to be used in existing light-water reactors.

10 While pioneered and then abandoned by **US** research labs, fuel pyroprocessing research at lab scale has continued in **South Korea**, which would like to begin larger scale demonstrations. Pyroprocessing technology is critical to fast reactors like GE-Hitachi's PRISM design, based on the Experimental Breeder Reactor (EBR-II) developed in the US by Argonne National Laboratory from 1964 to 1994, until Congress cut funding for the program.

Carbon Capture and Storage

Despite enormous efforts to the contrary, the world is becoming more, not less, dependent on coal, the most carbon-intensive large-scale power source.[25] This testifies to both poor countries' desire to transcend poverty and the powerful path dependencies that govern technological innovation. China already burns as much coal as the rest of the world combined. By 2030, the developing Asia-Pacific region, led by China, is expected to double its electricity demand to consume more electricity annually than all of the affluent OECD countries put together. The Asian Development Bank projects that 83 percent of these energy needs will be met with coal.[26]

Its commitment to greener growth notwithstanding, China is building out the energy system that will meet this tremendous demand in substantial part by exploiting coal, its largest and cheapest energy resource.[27] In the past decade alone, China has invested many hundreds of billions of dollars in its "coal-based quest for modernity."[28] The size of China's coal endowment and its need to exploit this resource to meet the needs of its people in the absence of other inexpensive, large-scale power options means that putting a dent in global emissions from the energy

---

[25] John McGarrity, "Coal to rival oil as dominant energy source by 2017: IEA," *Reuters* (18 Dec. 2012).

[26] Jacqueline Koch, "Learning from China: A Blueprint for the Future of Coal in Asia? An Interview with Armond Cohen," National Bureau of Asian Research (NBR), for the 2014 Pacific Energy Forum (21 Apr. 2014), available at: http://www.nbr.org/downloads/pdfs/eta/PEF_2014_Cohen_interview_04222014.pdf

[27] Jeffrey Ball, "The Battle in Our Trade War with China," *The New Republic* (21 Jan. 2014).

[28] Vaclav Smil, *Energy Transitions: History, Requirements, Prospects* (Santa Barbara, CA: Praeger, 2010), p. 126.

sector will depend on the construction of efficient coal power plants equipped with carbon capture and storage (CCS) in China.

The dependence on coal-based electricity generation shared by China and the United States also represents an opportunity for creating new models of international collaboration on energy innovation. To this end, the Clean Air Task Force's (CATF) Asia Project provides a platform for Western technology developers to collaborate with Asian partners.[29] Through workshops, conferences, and briefings in the United States and China, CATF helps bring about joint business ventures that leverage both countries' extensive experience with coal. For example, Chinese firms have estimated significantly lower costs for capturing emissions[30]—an example of the "China price" for energy innovation—and U.S. companies have experience with using carbon dioxide for enhanced oil recovery (EOR) and storage in geologic formations.

Partnerships that build on knowledge exchanges, such as the series of U.S.-China energy pacts,[31] and the potential financing offered by the New Development Bank (or BRICS bank),[32] decrease technology costs and speed deployment. Such efforts are also vital for bridging cultural and communication divides that have hindered effective international cooperation and have for decades been the undoing of the "technology transfer" model.

---

[29] Clean Air Task Force, CATF Asia Project, available at: http://www.catf.us/fossil/where/asia/

[30] Xi Liang and David Reiner, "How China can kick-start carbon capture and storage," *ChinaDialogue* (28 May 2013).

[31] Valerie Volcovici and Michael Martina, "U.S., China ink coal, clean energy deals but climate differences remain," *Reuters* (9 Jul. 2014).

[32] Keith Johnson, "Coke Brothers," *Foreign Policy* (22 Jul. 2014).

● OXY-FUEL COMBUSTION CAPTURE

1. Compostilla - **Switzerland**-based multinational Foster Wheeler is a major investor in this project in northwest **Spain**. The European Commission provided 180 million euros to the project as part of Europe's economic recovery strategy.

2. Datang Daqing - **France**-based Alstom and **China**-based Datang collaboration on CCS power plant demonstration in China.

● PRE-COMBUSTION CAPTURE (GASIFICATION)

3. TharPak - Consortium of **American** companies and multinationals financing a coal plant with CCS in the Thar region of **Pakistan**.

4. GreenGen - Peabody Energy (headquartered in **St. Louis, Missouri**) is a major investor in this northern **China** project.

5. Dongguan - KBR (headquartered in **Texas**) and the Southern Company (headquartered in **Atlanta, Georgia**) are providing the IGCC technology for this CCS project in southern **China**.

6. Kemper - Southern Co. and Shenhua signed a deal to share technology and knowledge related to the Kemper coal gasification plant in **Mississippi**. Work will include other government agencies and universities as well.

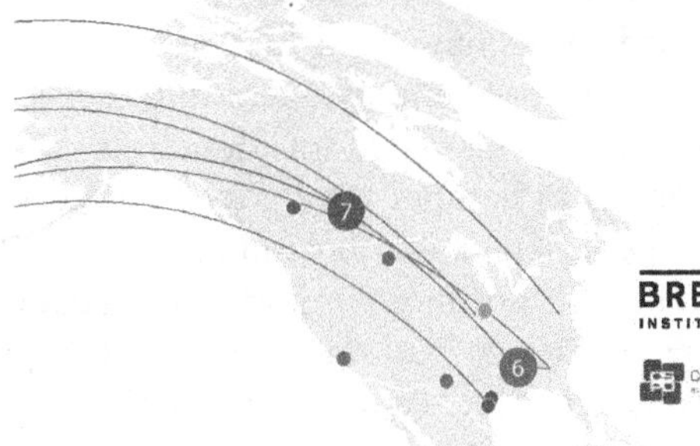

Carbon Capture and Storage Innovation

Governments and industry collaborate on carbon capture technologies with an aim towards producing power while capturing and storing emissions underground.

● POST-COMBUSTION CAPTURE

7 Boundary Dam - Saskatchewan Power Corporation and **Japan**-based Hitachi collaborated on one of the world's first and largest CCS power plants, that became active in fall of 2014.

8 Porto Tolle - **Norway**-based multinational Aker Solutions is a major investor in this project in northern **Italy**. The European Commission provided 100 million euros as part of Europe's economic recovery strategy.

9 Getica - The Global CCS Institute (headquartered in **Australia**) provided AU$2.5 million towards this project planned for operation in 2015 in **Romania**.

10 Taweelah - Masdar in partnership with BP.

11 **Korea** CCS 1&2 - are being developed by the Korea Carbon Capture and Sequestration research Center, which has an agreement with Scottish Carbon Capture and Sequestration (SCCS), the largest CCS R&D center in the **United Kingdom**.

The climate and development communities have for the most part refused to acknowledge that the world is irrevocably committed to fossil energy, including coal, for the foreseeable future.[33] This state of denial is self-defeating and continues to hamper efforts to drive down the costs of CCS through demonstration and operation not only in the United States,[34] but more importantly in countries that require assistance in building out their energy sectors. For example, relieving Pakistan of its crippling energy poverty by exploiting the country's Thar coal deposits—a project which integrates a range of advanced carbon capture technologies—has received no support from the U.S. Agency for International Development (USAID), despite the fact that USAID considers energy sector development in Pakistan a "top assistance priority."[35] Development of mining and power generation in the Thar region has finally begun with the backing of Chinese

---

[33] On this refusal, see, for example, the open letter from Friends of the Earth Africa to President Barack Obama regarding the Power Africa initiative and the Electrify Africa Act of 2013 (subject line: "Leave the oil in the soil; leave the coal in the hole"), available at: http://libcloud.s3.amazonaws.com/93/a5/b/3320/11-8-13_Power_Africa_lett_FINAL.pdf. In the North American context, see this 16 Jan. 2014 open letter from the Sierra Club and 17 other environmental groups to President Obama regarding his pragmatic "all of the above" energy strategy, available at: http://action.sierraclub.org/site/DocServer/All_of_the_Above_letter_Jan_16_FINAL_corrected.pdf?docID=14881.

[34] International Energy Agency, *21st Century Coal: Advanced Technology and Global Energy Solution*, Report by the IEA Coal Industry Advisory Board (Paris, France: OECD/IEA, 2013).

[35] Steven Michael Carpenter, "A Multi-Quadrant Analysis of Two International Climate Mitigation Projects," master's thesis, Antioch University (Aug. 2012); Richard Silver, "U.S. Officials Visit Hyderabad Focusing on Education and Energy," Consulate General of the United States. (June 2013).

banks, but whether the project will ultimately incorporate CCS is unclear.[36] Regardless of the merits of any particular plan, a soft-energy framework restricted to non-fossil energy impedes progress in CCS, despite the fact that the world will depend on coal for decades to come.

Solar Photovoltaics

The last several decades have witnessed a remarkable reduction in the cost of solar photovoltaics (PV), declining by a factor of 100 over the past 50 years.[37] Extremely high early costs for solar PV found a niche market in satellite application and experienced marked cost improvements thereafter, most notably through R&D improvements in electrical conversion efficiency of PV systems, manufacturing scale, and technological "learning" in response to local markets and public deployment policies.[38] Solar experts have emphasized the "chain-linked innovation" of

---

[36] Ijaz Kakakhel, "Thar coal project: Chinese banks reluctant to accept sovereign guarantees," *Daily Times* (26 Oct. 2014), available at: http://www.dailytimes.com.pk/national/26-Oct-2014/thar-coal-project-chinese-banks-reluctant-to-accept-sovereign-guarantees

[37] Gregory Nemet, "Solar Photovoltaics: Multiple Drivers of Technology Improvement. Historical Case Studies of Energy Technology Innovation," Ch. 24 in *The Global Energy Assessment*, A. Grübler, F. Aguayo, K. S. Gallagher, M. Hekkert, K. Jiang, L. Mytelka, L. Neij, G. Nemet, and C. Wilson, eds. (Cambridge, UK: Cambridge University Press, 2012).

[38] T. Surek, "Progress in U.S. Photovoltaics: Looking back 30 years and looking ahead another 20," 3rd World Conference on Photovoltaic Energy Conversion, Osaka, Japan (2003); M. A. Green, "Silicon photovoltaic modules: a brief history of the first 30 years," *Progress in Photovoltaics: Research and Applications*, Vol. 13, No. 5 (2005): pp. 447-455.

R&D, production support, and market formation that, combined, enabled solar PV to reach its current status.[39]

The recent dramatic cost declines in solar PV cells and modules are the result of interacting international technology policies: Western deployment regimes paired with aggressive Chinese industrial policy and pursuit of solar PV manufacturing dominance. Policies like the U.S. federal investment tax credit (ITC) for solar and European feed-in tariffs provided the demand that Chinese solar manufacturers sought to supply through aggressive national and regional state subsidization of solar production capacity. However, the policies that interacted to drive these cost declines also resulted in international solar trade wars, as Western governments—including Germany and the United States—accused Asian solar producers of flooding the market to gain an unfair competitive edge. How this supply-side dispute will be resolved is not yet clear.[40] And whether countries like the United States and Japan experience a slowdown in the deployment of solar PV similar to that of Spain and Germany also remains to be seen.[41]

In the meantime, more and more emerging economies are taking advantage of the price decline in PV by foster-

---

[39] Kelly Sims Gallagher, Arnulf Grübler, Laura Kuhl, Gregory Nemet, and Charlie Wilson, "The Energy Technology Innovation System," *Annual Review of Environment and Resources*, Vol. 37 (2012): pp. 137-162.

[40] Matthew Stepp and Robert D. Atkinson, "Green Mercantilism: Threat to the Clean Energy Economy," Information Technology & Innovation Foundation (7 June 2012), available at: http://www.itif.org/publications/green-mercantilism-threat-clean-energy-economy

[41] Craig Morris, "German PV market continues shrinking," *Renewables International* (5 Oct. 2014); Paul Voosen, "Spain's Solar Market Crash Offers a Cautionary Tale About Feed-In Tariffs," *The New York Times* (18 Aug. 2009).

ing solar markets of their own. From Asia to the Middle East to Latin America, the market for solar plants is becoming similar to the market for upstream solar manufacturing: "truly global."[42] The leading manufactures of global PV—including the United States' First Solar, China's Yingli and Trina, Canada's Canadian Solar, Germany's SolarWorld, Japan's Sharp Electronics, and others—are increasingly exporting their panels for us in large solar farms in emerging economies. Development and finance of new solar capacity is increasingly international as well, with firms like U.S.-based SunEdison developing a 100-megawatt plant in Chile,[43] a German development bank financing a 150-megawatt plant in India,[44] and two Russian banks financing a 105-megawatt plant in Ukraine.[45]

Solar PV's progress is substantial and ongoing. But as solar expert Greg Nemet described in a 2012 case study charting PV's decades of progress, "despite this achievement, the technology remains too expensive compared to existing electricity sources, such that widespread deployment depends on substantial future improvements."[46] To date, the increasingly globalized deployment of solar PV still depends largely on concerted government efforts. The International Energy Agency expects that solar PV will contribute a relatively minor portion of global electricity production in 2050 without major improvements made to

---

[42] Nemet, "Solar Photovoltaics."

[43] Michelle DiFrangia, "South America's Largest PV Plant Officially Introduced," *Solar Power World Online* (11 June 2014).

[44] Aparna Pallavi, "Sakri solar power plant may not be shifted after all," *Down to Earth* (9 Nov. 2011).

[45] Clean Energy Action Project, "Perovo Solar Power Station: Case Study."

[46] Nemet, "Solar Photovoltaics."

MANUFACTURING

China dominates solar PV production, with 7 of the 10 largest manufacturers in the world.

POWER PLANTS

Amanecer - This 100-megawatt plant is the largest solar PV plant in **Latin America**.

Gujarat - This project in **India** will be Asia's largest solar power park hub.

Longyanxia Dam Solar Park - This **Chinese** plant will be the world's largest solar-hydro power station.

Topaz Solar Farm - At 550 megawatts, this **San Luis Obispo** plant is the largest photovoltaic plant in the world.

Latin America: **Brazil, Argentina**, and other Latin American countries are aggressively jumping into the solar PV market, with a total of 6,000 megawatts of solar under development.

Saudi Arabia has planned 16,000 megawatts of solar PV capacity to be deployed over the next 20 years, as well as 25,000 megawatts of solar thermal.

German renewables firm Juwi is developing several large PV power plants in **South Africa** under an MOU between the German and South African governments.

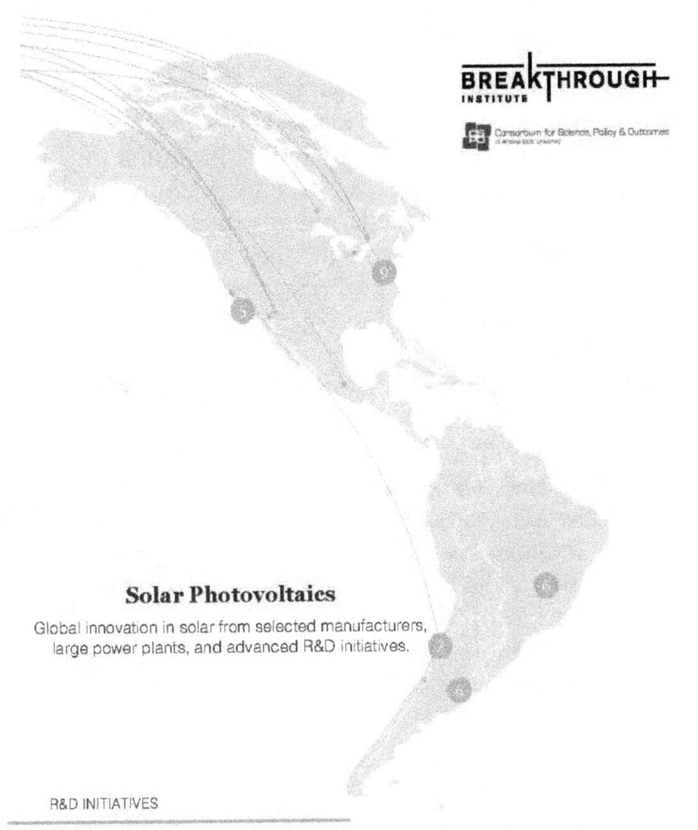

R&D INITIATIVES

9. Solar Energy Center - A collaboration between the **US** Department of Energy National Renewable Energy Laboratory (NREL) and the **Indian** Solar Energy Center (SEC) to test new manufacturing techniques for thin-film solar PV.

10. ArtESun - The goal of ArtESun is to develop advanced organic solar cell technologies. This is a collaboration among top research groups and industries within **Belgium**, **Germany**, **the UK**, **Spain**, **France**, **Finland**, and **Canada**.

11. Solar & Photovoltaics Engineering Research Center (SPERC) - In spring of 2014 a 5-year MOU was signed between ZSW in **Germany** and KAUST in **Saudi Arabia** to develop advanced thin-film solar technology.

12. IEA Photovoltaic Power Systems Programme - One of the International Energy Agency's several R&D Agreements, established in 1993, dedicated to pursuing advanced solar technologies.

a range of technologies, including storage and transmission, as well as to business models and policy.[47]

Fortunately, there are encouraging efforts and investments being made towards innovation in advanced and next-generation solar PV technologies, including organic PV and thin-film. Several countries are pursuing a brand of solar industrial policy via public-private partnerships in advanced solar PV innovation. These include a partnership between Merck and the German government in pursuit of breakthroughs in organic PV,[48] JA Solar's partnership with the Chinese Academy of Sciences,[49] and the Brazilian Technology System's (SIBRATEC) investments in solar innovation initiatives.[50] There are also many budding internationally collaborative efforts being made towards next-generation solar innovation. Key among these is the International Energy Agency's 29-member Photovoltaic Power Systems Programme, which includes the United States, China, Germany, Malaysia, Mexico, Turkey, and others.[51] Other promising activity includes a memorandum of understanding in thin-film research between research institutes in Germany and Sau-

---

[47] Alex Trembath, "Not Solar Fast: Rapid Expansion of Solar Depends on Massive Subsidies and High Carbon Price," Breakthrough Institute (1 Oct. 2014).

[48] Ian Clover, "German government launches €16 million organic PV R&D project," *PV Magazine* (13 Nov. 2013).

[49] Heather Reidy, "Cutting-edge PV centre set for China," *Renewable Energy Technology* (3 May 2012).

[50] International Energy Initiative, "Workshop Innovation for the Establishment of Photovoltaic Energy Sector in Brazil" (31 Mar. 2011) available at: http://iei-la.org/workshop-innovation-for-the-establishment-of-photovoltaic-solar-energy-sector-in-brazil/

[51] International Energy Agency, "IEA Photovoltaic Power Systems Programme," available at: http://www.iea-pvps.org/index.php?id=4 - c38

di Arabia,[52] and a partnership between the U.S. Department of Energy and India's Solar Energy Center.[53]

[52] King Abdullah University of Science and Technology, "ZSW joins forces with KAUST in thin-film photovoltaic technology" (1 May 2014), available at: http://www.kaust.edu.sa/latest-stories/zsw-signs-alliance-with-kaust-in-thin-film-photovoltaic-research.html

[53] U.S. Department of Energy, "International Team," U.S. DOE Office of Energy Efficiency & Renewable Energy, available at: http://energy.gov/eere/about-us/international-team

8

CLEAN ENERGY AS PUBLIC GOOD

In *The Post-American World,* Fareed Zakaria observed that the rise of large developing powers like the BRICS (Brazil, Russia, India, China, and South Africa) augured the relative, but not absolute, decline of the West.[1] Many wealthy developed nations, with their strong scientific and engineering institutions, have an interest in engaging "the rise of the rest" for geopolitical, economic, and environmental reasons. This line of reasoning applies well to energy innovation. Just as U.S.-pioneered solar and nuclear energy technologies are benefitting China today, next-generation versions of those technologies could benefit the United States—and the world—in the future.

Clean energy innovation should thus be recognized as a public good and a shared responsibility. Energy-climate innovations should be considered public goods similar to national defense, public health, adequate food supplies, or a safe air transportation network, where governments routinely invest billions of dollars to advance specific technologies that solve particular problems. Treating this kind of innovation as a public good offers new avenues for public investment in promising energy technologies and technology portfolios.

---

[1] Fareed Zakaria, *The Post-American World* (New York, NY: W.W. Norton & Company, 2009).

There is a long history of effective international collaboration on innovation for global public goods. One of the most well-known and most successful was the creation and funding of international agricultural research centers by the philanthropies and governments behind the Green Revolution.[2] More recently, developed and developing countries, along with philanthropy and the private sector, have worked together in novel partnerships to accelerate biomedical innovations to address health challenges ranging from AIDS to malaria to tuberculosis.[3] Like innovation collaborations around agriculture and health, innovations in energy technologies benefit all nations involved and thus can be justified by states and societies as worthy of shared investment.

The influence of the low-energy development model has distracted attention from the urgent need to mobilize these public-private energy innovation projects. This is the case not only for renewables, but also for CCS and nuclear power, which have historically failed the litmus test of environmental correctness.[4] But a consensus of experts, Nobel Prize winners, and respected international leaders have over the last decade called for a greater public and private sector commitment to improving low- and zero-carbon technologies.[5] Bill Gates and others have empha-

---

[2] Robert E. Evenson and Douglas Gollin, "Assessing the impact of the Green Revolution, 1960 to 2000," *Science*, Vol. 300, No. 5620 (2003): pp. 758-762.

[3] John Clemens, Jan Holmgren, Stefan H. E. Kaufmann, and Alberto Mantovani, "Ten years of the Global Alliance for Vaccines and Immunization: Challenges and Progress," *Nature Immunology*, Vol. 11, No. 12 (2010): p. 1069.

[4] Armond Cohen, "Recent Expert Reports: Diverse Zero Carbon Options Needed to Manage Climate," Clean Air Task Force (21 Sept. 2014).

[5] See, for example: "Report to the President on Accelerating the Pace of Change in Energy Technologies Through an Integrated

sized the multiple benefits from energy innovation. "If you gave me only one wish for the next 50 years—I could pick who is president, I could pick a vaccine ... or I could pick [an energy technology] that's half the cost with no carbon emissions—this is the wish I would pick," Gates said. "This is the one with the greatest impact."[6]

Due to rising technological complexity and the cost of innovation, a strong public role is essential.[7] While in 1858, two industrialists could break rock and produce the first oil in North America, today even the richest individuals could not pioneer a new nuclear technology without the support of the Chinese state.[8] New energy technologies—whether solar, batteries, nuclear, or biofuels—require increasingly specialized knowledge, integration with ever larger and more complex infrastructures, and industrial capacities that span many sectors, nations, and institutions. In this context, the role of the state may be rising. Today's cheap solar panels were the result of massive public Chinese investments at both the national and

---

Federal Energy Policy" from the President's Council of Advisors on Science and Technology (PCAST), which calls for $12 billion in annual federal funding for energy research, development and demonstration (RD&D); "A Business Plan for America's Energy Future" from the American Energy Innovation Council (AEIC), which calls for $16 billion in annual federal funding for energy RD&D; a letter from 34 Nobel Laureates calling for $15 billion in annual federal funding for clean energy R&D; a letter signed by the Association of Public and Land-grant Universities (APLU) and the Association of American Universities (AAU) calling for $15 billion in annual federal funding for clean energy R&D.

[6] Bill Gates, "Innovating to Zero!" TED Talks (Feb. 2010).

[7] Jesse Jenkins, Devon Swezey, and Yael Borofsky, "Where Good Technologies Come From: Case Studies in American Innovation," Breakthrough Institute (5 Dec. 2010).

[8] Michael del Castillo, "Bill Gates' TerraPower races to generate safe power, in China," *Upstart Business Journal* (26 Sept. 2013).

local levels. Elon Musk required a half-billion dollars in taxpayer subsidies to achieve the highly regarded (yet still extremely expensive) Tesla Model S electric vehicle, not to mention the billions of dollars of investments by Japanese and American governments in the necessary pre-competitive research and development over the previous 20 years.[9]

Governments have multiple roles to play. Probably the most notable aspect of fracking's development was the close collaboration between public institutions and private firms. Rather than keeping the private sector at arm's length—as is often considered best by economists wary of corporate capture—the U.S. Department of Energy (DOE) developed gas research programs in which gas companies were explicitly asked to contribute.[10] One such program, the Eastern Gas Shales Research Program, is credited for pioneering horizontal drilling for shale gas. DOE was also responsible for approving the funding and research efforts of the Gas Research Institute (GRI), an industry consortium tasked with developing new energy technologies. Working closely with the private sector, and providing technical expertise that complemented the practical experience of private sector firms, the state played a decisive role in developing the technology without spending vast sums of money.[11]

On the face of it, the fracking case is yet another challenge to the common presumption that the role of the state

---

[9] Michael Shellenberger and Ted Nordhaus, "The Myth of the Lone Innovator," *The New Republic* (30 May 2013).

[10] Jim Manzi, "The New American System," *National Affairs*, Vol. 19 (2014).

[11] Jason Burwen and Jane Flegal, "Case Studies on the Government's Role in Energy Technology Innovation: Unconventional Gas Exploration and Production," American Energy Innovation Council (Mar. 2013).

in energy innovation is to fund "basic science" and stand back while the private sector develops the technologies.[12] In this case the government's most important role seems to have been as network builder, coordinating research and development with multiple actors in the field, not in laboratories unconcerned with economic considerations under the guise of basic or "pure" science.

Unfortunately, the lessons of the fracking revolution for clean energy innovation remain largely unknown to policymakers. While local environmental consequences of natural gas production must be addressed, these concerns have obscured the most significant aspects of the shale revolution. Namely, that public-private collaboration over three decades has rapidly reduced U.S. emissions and may prove to reduce carbon intensity globally.[13] This collaboration should be a model for American philanthropies as well as donor countries, which to date have been overly focused on the low-energy paradigm and expanding the small-scale deployment of largely intermittent renewables.

Whether or not nations increase their investment in innovation, they can still pursue visionary energy innovation collaborations, as the United States' collaboration with China on nuclear demonstrates. Recognizing that the United States was unlikely to demonstrate advanced salt-cooled nuclear reactors, the Oak Ridge National Laboratory agreed to share this information with the Chinese government in 2010.[14] A case can be made that the DOE and

---

[12] Michael Shellenberger and Ted Nordhaus, "Reinventing Libertarianism: Jim Manzi and the New Conservative Case for Innovation," Breakthrough Institute (22 Apr. 2014).

[13] Alex Trembath, Ted Nordhaus, Michael Shellenberger, and Max Luke, "Coal Killer: How Natural Gas Fuels the Clean Energy Revolution," Breakthrough Institute (June 2013).

[14] David Lague and Charlie Zhu. "The U.S. government lab behind China's nuclear power push," *Reuters* (20 Dec. 2013).

U.S. firms should compete with the Chinese government and Chinese firms to create the best new nuclear reactor, and step up both public and private efforts. But barring a radical and unlikely increase in U.S. nuclear innovation efforts, China will lead the global race for both development and deployment of advanced nuclear reactors.

To catalyze, augment, and legitimate such public support, philanthropy can play a pivotal role. In the 1940s, the Rockefeller and Ford Foundations made investments in international agricultural innovation that were pivotal in catalyzing the Green Revolution.[15] More recently, the Bill and Melinda Gates Foundation (along with several others) has been crucial in accelerating innovation to meet urgent global public health challenges.[16] Philanthropies and the organizations they fund have long time horizons and the ability to set ambitious programs that transcend national boundaries, technology classes, and economic sectors, without succumbing to short-term political pressures.

A good first step for philanthropies would be to develop comprehensive "maps" of important programs, public and private, across the globe as a basis for strategic investments in partnerships that can make crucial contributions to scalable, inexpensive, low-carbon energy in the near- to medium-term. As our brief case studies are meant to show, partnerships that bring together different innovation capacities within and between nations, that foster trust between partners, and that reflect and emerge from

---

[15] John H. Perkins, "The Rockefeller Foundation and the green revolution, 1941–1956," *Agriculture and Human Values*, Vol. 7, Nos. 3-4 (1990): pp. 6-18.

[16] See, for example: Selcuk Özgediz, "Organisation and management of the CGIAR system: a review," *Public Administration and Development*, Vol. 13, No. 3 (1993): pp. 217-231; and Mitch Renkow and Derek Byerlee, "The impacts of CGIAR research: A review of recent evidence," *Food Policy*, Vol. 35 (2010): pp. 391-402.

the realities of existing energy systems and markets, can be key to accelerated innovation. And such partnerships can be fostered without enormous new government investments.

Behind two decades of political stalemate on climate are national and economic interests that cannot be transcended as long as the priority is to reduce energy consumption and pay more for energy. When the focus is reversed, and governments, industry, and philanthropies collaborate to provide the global public good of abundant, clean, cheap energy, climate policy will win far higher levels of public support. Embracing high-energy innovation is the best way to address our shared energy and climate challenges.

ADAPTATION

FOR A HIGH-ENERGY PLANET

— A CLIMATE PRAGMATISM PROJECT —

NETRA CHHETRI, JASON LLOYD, TED NORDHAUS, ROGER PIELKE JR., JOYASHREE ROY, DANIEL SAREWITZ, MICHAEL SHELLENBERGER, PETER TEAGUE, AND ALEX TREMBATH

Consortium for Science, Policy & Outcomes

MARCH 2016

BREAKTHROUGH INSTITUTE

9

FAILING TO MITIGATE, FAILING TO ADAPT

For almost a generation, adaptation in response to a changing global climate has been the neglected stepchild of climate policy. In at least the first decade and a half after the establishment of the United Nations Framework Convention on Climate Change (UNFCCC) in 1992, adaptation was viewed negatively by many of the architects of the process, even as developing countries pushed for greater investment in adaptation efforts. Central to international efforts to address the climate challenge was mitigating through global greenhouse gas (GHG) reductions. Adaptation was perceived as a distraction to these efforts, since, as the UNFCCC asserts, "At the very heart of the response to climate change … lies the need to reduce emissions."[1]

Even as adaptation has more recently gained mainstream acceptance as an unavoidable response to rising global temperatures, it continues to be a sideshow to the main event in the international climate arena: efforts to establish limits on GHG emissions. This single-minded approach has disappointed in myriad ways. Climate poli-

---

[1] "Background on the UNFCCC: The international response to climate change," available at: https://unfccc.int/essential_background/items/6031.php

cies with an overriding concern toward mitigation have tended to foster political gridlock, encourage alarmist rhetoric, and entrench inequities in, for example, access to modern energy.

While demonstrably failing to achieve its mitigation targets, the UNFCCC has also constrained international adaptation efforts by narrowly defining adaptation in terms of responses to changes in climate that are attributable to human causes.[2] Such a definition effectively ignores the vulnerability to a naturally capricious climate that much of the global population already faces, irrespective of any additional forcing attributable to anthropogenic GHG emissions. At the same time, the UNFCCC's definition can in practice predicate adaptation action on the ability to distinguish climate impacts that result from human-caused global warming from those that do not—a near-absurd distinction.[3]

The present international framework for climate adaptation has been forced into the procrustean bed of climate change mitigation policy. Thus it has missed enormous opportunities for effective action to reduce human suffering due to climate variability and weather disasters, and to lay a stable foundation for cooperative international efforts to address *both* climate adaptation and mitigation. This foundation, we argue, would be greatly strengthened by refocusing attention on broad-based adaptation efforts—particularly those that complement inclusive, low-carbon economic growth strategies. Now is the time to pragmatically rethink our approach to adaptation by con-

---

[2] Roger A. Pielke Jr., "Misdefining 'climate change': consequences for science and action," *Environmental Science & Policy*, Vol. 8 (2005): pp. 548-561.

[3] Roger Pielke Jr., Gwyn Prins, Steve Rayner, and Daniel Sarewitz, "Lifting the taboo on adaptation," *Nature*, Vol. 445 (8 Feb. 2007): pp. 597-598.

necting it to a diversity of values and aspirations—many of which have little identifiable connection to anthropogenic climate change.

A crowd calls for mitigation measures to deal with climate change near the Washington Monument. *(Photo credit: Jmcdaid.)*

We propose as a concrete goal for adaptation policies the progressive and continual reduction of the human toll (measured as average number of deaths each year) from natural disasters, including those disasters that will be affected or worsened by a changing climate. Tangible and universally appealing, this goal offers the potential for leveraging actions whose benefits are measurable in the short term. It opens a tremendous number of options for action—humans are nothing if not ingenious in adapting to a dynamic and often hazardous environment[4]—and can

[4] See, for example: Anupa Ghosh and Joyashree Roy, "Synthesis Report on Floods and Droughts," *Science and Culture*, Vol. 71, Nos. 7-8 (2005): pp. 285-287; Anupa Ghosh and Joyashree Roy, "Coping with Extreme Climatic Events: Analysis of Household and Community Responses from Selected Hotspots in India," *Science and Culture*, Special Issue on Flood Disaster, Risk Reduction in Asia, Vol. 72, Nos. 1-2 (2006): pp. 23-31.

provide a politically attractive and morally appealing vision for development initiatives such as the United Nations' ambitious Millennium Development Goals and their successors, the Sustainable Development Goals.

Progress toward systematically reducing deaths from natural disasters is measurable using the most tangible of yardsticks—the saving of lives—with direct payoffs in the present. The objective may seem formidable on a planet with a growing population, and requires shifting the conversation about climate adaptation to focus on peoples' well-being, equity, and livelihoods. But it is certainly no more ambitious than the radical economic and technological transformations required by drastic emissions reductions. And it is one that many societies in different parts of the world have actually succeeded at achieving.

Adoption of such a goal can open up climate policy to new, politically invigorating options and opportunities. Above all, unshackling adaptation from mitigation-centered institutions like the UNFCCC requires understanding adaptation as a central component of socioeconomic development. Shifting to adaptation-focused objectives for the many funding mechanisms, policy levers, and lending institutions that currently operate in poor countries under the rubric of climate mitigation—some which enable and some which inhibit truly sustainable development—can move these efforts in a more pragmatic and coherent direction, one that directly improves people's lives. Adaptation provides an agenda that policymakers can implement at the national and local level, shaping decisions about urban and community development, land use and resource management, hazard insurance, building codes, evacuation, and recovery efforts that impact how communities prepare for and experience crises.[5]

In places like Ningbo, where three large rivers converge, China is undertaking substantial measures to reduce vulnerability to flash flooding. *(Photo credit: Jiong Sheng.)*

International adaptation efforts can also drive innovation in critical areas like food production, public health, and energy provision. Making societies more resilient increases their ability to innovate and mitigate around other challenges, notably the increasing concentration of greenhouse gases in the atmosphere. And since adaptation addresses universal concerns about safety and well-being, an agenda focused on improving adaptation to climate variability does not demand agreement on climate science or particular mitigation measures: it can embrace and benefit from a healthy political pluralism that is unavailable to today's rigid mitigation-oriented policy structure. This

[5] Roger A. Pielke Jr. and Daniel Sarewitz, "Bringing Society Back into the Climate Debate," *Population and Environment*, Vol. 26, No. 3 (Jan. 2005): p. 258.

lack of inclusiveness has been, in part, responsible for its undoing.[6]

As a political matter, adaptation efforts as broadly conceived (essentially, "adaptation to climate" rather than "adaptation to climate *change*," although the latter may become more important in the future) reinforces a general political consensus over the responsibilities of government to provide the infrastructure, policies, and services necessary for thriving, and thus can counter the corrosive effect that climate change debates have sometimes had on democratic politics.[7] "Low-regret" and "win-win" efforts to directly improve and protect peoples' lives while supporting necessary mitigation efforts and other co-benefits offer attainable objectives for creating a more prosperous and resilient world that resonates with a diversity of values.

In the chapters that follow, we evaluate opportunities to increase climate resilience through standard concepts used to assess and mitigate risk; consider successful adaptations in a variety of global contexts in search of key les-

---

[6] Rob Atkinson, Netra Chhetri, Joshua Freed, Isabel Galiana, Christopher Green, Steven Hayward, Jesse Jenkins, Elizabeth Malone, Ted Nordhaus, Roger Pielke Jr., Gwyn Prins, Steve Rayner, Daniel Sarewitz, and Michael Shellenberger, *Climate Pragmatism: Innovation, Resilience, and No Regrets* (Oakland, CA: Breakthrough Institute, Jul. 2011).

[7] Nico Stehr, "Democracy is not an inconvenience," *Nature*, Vol. 525 (24 Sep. 2015): pp. 449-450; Steven F. Hayward, "Conservatism and Climate Science," *Issues in Science and Technology*, Vol. 30, No. 3 (Spring 2014): pp. 52-57. See also: Douglas R. Weiner, "Demythologizing Environmentalism," *Journal of the History of Biology*, Vol. 25, No. 3 (Autumn 1992): pp. 385-411. For a broader discussion of scientific knowledge and its connection to authoritarianism, see: James C. Scott, *Seeing Like a State: How Certain Schemes to Improve the Human Condition Have Failed* (New Haven, CT: Yale University Press, 1998), especially pp. 88-90.

sons for international climate adaptation; and consider what an alternative framework that takes effective adaptation as its primary objective might look like.

10

DISASTER RISK & CLIMATE CHANGE

Losses caused by disasters are the result of three factors: the hazard, exposure, and vulnerability.[1] **Hazards** are extreme events such as floods, hurricanes, or droughts. The origin, frequency, or severity of a hazard may have a human element that some risk-mitigation efforts attempt to reduce. Reforesting denuded hillsides, for example, can prevent landslides; dredging and widening river channels can help prevent floods. Similarly, mitigation-oriented climate policies are justified by the idea that a future of more intense or more frequent storms, droughts, and temperature extremes can be avoided through emissions reductions.

Exposure reflects the human development that is subject to the adverse effects of hazards. We live in a dynam-

---

[1] Reinhard Mechler and Laurens M. Bouwer, ""Understanding trends and projections of disaster losses and climate change: is vulnerability the missing link?" *Climatic Change*, DOI 10.1007/s10584-014-1141-0; Omar-Dario Cardona, M.K. van Aalst, J. Birkmann, M. Fordham, G. McGregor, R. Perez, R.S. Pulwarty, E.L.F. Schipper, and B.T. Sinh, "Determinants of Risk: Exposure and Vulnerability," Ch. 2 in *Managing the Risks of Extreme Events and Disasters to Advance Climate Change Adaptation*, C.B. Field, et al., eds. A Special Report of Working Groups I and II of the Intergovernmental Panel on Climate Change (IPCC) (New York, NY: Cambridge University Press, 2012), pp. 65-108.

ic, unpredictable world, and there are few if any places in which we are not exposed to some natural disaster or other. Unfortunately, much of the world's wealth and population are concentrated in areas especially prone to climate risk, such as coastal areas and flood plains. Some interventions can reduce or prevent exposure, like smart land-use planning or building codes.

Vulnerability describes the susceptibility of people, their livelihoods, or their property to suffer negative effects when exposed to a hazard. A complex set of factors plays a role in whether or not people are vulnerable to climate hazards. Socioeconomic development is one of the most significant of these factors. Prosperous, well-governed, democratic communities have more resources and capacity to devote to protecting themselves from current hazards and adapting to deal with new ones.[2] This is why we tie our adaptation agenda so closely to development processes, such as improved energy access and greater innovation capacity. Without that development, poor communities can lack the capacity to manage their vulnerability to hazards, regardless of how well the world may progress in slowing climate change.

Understanding the Climate Hazard

The central tenet of climate change mitigation policies has been that reducing GHG emissions today will reduce the consequences of future climate hazards by preventing human-caused increases in the magnitude, frequency, or duration of hazards like hurricanes, floods, sea level rise, and droughts. The idea of reducing the hazards associated with anthropogenic climate change by preventing them in

---

[2] Matthew E. Kahn, "The Death Toll from Natural Disasters: The Role of Income, Geography, and Institutions," *The Review of Economics and Statistics* Vol. 87, No. 2 (2005): pp. 271-284.

the first place may have made some notional sense as the UNFCCC was being formulated in the late 1980s and early 1990s. But the overwhelming evidence from two decades of peer-reviewed research since suggests that such hopes are an extraordinarily weak foundation for addressing the challenges of climate change that society will face this century.

One reason is that losses from climate hazards have little to do with changes to the climate. The increased toll of human suffering, both in terms of casualties and property damages, from climate-related natural disasters over the past century is overwhelmingly a function of changes in exposure and vulnerability, not increasing hazards due to climate change.[3] The dramatic rise in disaster losses across the globe due to these other factors exceeds increases associated with hazard intensity by an order of magnitude. This contrast will likely continue to be the case for many decades, if not centuries, to come.

In the special report on Managing the Risks of Extreme Events and Disasters to Advance Climate Change Adaptation (SREX), the Intergovernmental Panel on Climate Change (IPCC) observes that "[m]ost studies of long-term disaster loss records attribute these increases in losses to increasing exposure of people and assets in at-risk areas ... and to underlying societal trends—demographic, economic, political, and social—that shape vulnerability to impacts."[4] Likewise, in the latest IPCC Assessment Report (AR5), the Second Working Group concludes: "Economic

---

[3] Roger Pielke Jr., *The Rightful Place of Science: Disasters and Climate Change* (Tempe, AZ and Washington, DC: Consortium for Science, Policy & Outcomes, 2014).

[4] Virginia Murray and Kristie L. Ebi, "IPCC Special Report on Managing the Risks of Extreme Events and Disasters to Advance Climate Change Adaptation (SREX)," *Journal of Epidemiology and Community Health*, Vol. 66, No. 9 (2012): pp. 759-760.

growth, including greater concentrations of people and wealth in periled areas and rising insurance penetration, is the most important driver of increasing losses."[5] The authors note that "the worldwide burden of human ill-health from climate change is relatively small compared with the effects of other stressors and is not well quantified."[6]

Much of the world's property and people are concentrated in places that are exposed to hazards like hurricanes, illustrated here by the damage from Hurricane Sandy in Mantoloking, NJ. *(Photo credit: Mark C. Olsen.)*

---

[5] IPCC, *Climate Change 2014: Impacts, Adaptation, and Vulnerability*. Part A: Global and Sectoral Aspects. Contribution of Working Group II to the Fifth Assessment Report of the Intergovernmental Panel on Climate Change, C.B. Field, V.R. Barros, D.J. Dokken, K.J. Mach, M.D. Mastrandrea, T.E. Bilir, M. Chatterjee, K.L. Ebi, Y.O. Estrada, R.C. Genova, B. Girma, E.S. Kissel, A.N. Levy, S. MacCracken, P.R. Mastrandrea, and L.L. White, eds. (Cambridge, UK: Cambridge University Press, 2014), p. 680.

[6] IPCC, *Climate Change 2014: Synthesis Report*. Contribution of Working Groups I, II, and III to the Fifth Assessment Report of the Intergovernmental Panel on Climate Change, R.K. Pachauri and L.A. Meyer, eds. (Geneva, Switzerland: IPCC, 2014), p. 51.

Another reason that mitigation policy can have little impact on climate hazards for the foreseeable future is that, even if wildly successful, the world is stuck with at least some warming and the consequent impacts. Even dramatic reductions in emissions today will not significantly manifest in terms of either global temperature trends or sea level rise until late this century or early in the next.[7] The IPCC estimates that even if emissions had stabilized in the year 2000, preceding emissions would have already "committed" the Earth to another half-degree Celsius of warming by 2100.[8] Other studies suggest that due to atmospheric-oceanic dynamics, surface temperature warming could continue for several centuries even if emissions stop today.[9] In addition, climate models predict little impact on sea level rise between radical emissions reductions and business-as-usual scenarios.[10]

These points do not minimize the importance of emissions reductions. Temperature increases of as much as 4 to 6 degrees Celsius over the next several centuries present unknowable and perhaps dire risks to human societies.

---

[7] Tom M. L. Wigley, *The Science of Climate Change: Global and U.S. Perspectives* (Arlington, VA: Pew Center on Global Climate Change, 1999).

[8] IPCC, *Climate Change 2007: The Physical Science Basis*, Contribution of Working Group I to the Fourth Assessment Report of the Intergovernmental Panel on Climate Change, S. Solomon, D. Qin, M. Manning, Z. Chen, M. Marquis, K.B. Averyt, M.Tignor and H.L. Miller, eds. (Cambridge, UK: Cambridge University Press, 2007).

[9] Thomas Lukas Frölicher, Michael Winton, and Jorge Louis Sarmiento, "Continued global warming after CO_2 emissions stoppage," *Nature Climate Change*, Vol. 4, No. 1 (2014): pp. 40-44.

[10] Tom M. L. Wigley, "Intermediate Radiative Forcing Targets and Sea Level Stabilization," University of Adelaide and National Center for Atmospheric Research (NCAR) Review Paper (Jan. 2015).

We are not yet able to confidently predict the nature and scale of those risks, or the adaptive capacity of future human societies. Effective measures to reduce emissions in order to limit the magnitude of future climate change are hence warranted and valuable. But justifying those measures in terms of mitigating climate hazards now or in the near future is not only unwarranted, it has blinded climate policymakers and advocates to the opportunities for human development offered by innovation-focused strategies (particularly in energy), as we discussed in previous chapters.

Successfully working to reduce GHG emissions now may reduce the intensity and frequency of some climate hazards in the future. In the meantime, hurricanes and typhoons will ravage coastlines, severe droughts will imperil farmers and water supplies, and floods will regularly sweep away people, homes, and livelihoods. This is to say nothing of tropical diseases, wildlife habitat loss, the collapse of fisheries, soil degradation, air and water pollution, and the many other difficulties that currently confront populations worldwide.[11] Climate mitigation efforts alone can do little to reduce present vulnerability to these disasters, and can perversely expose fragile populations to greater risk. This is precisely what happens when the narrow objective of limiting carbon emissions takes precedence over socioeconomic development that is environmentally sound and socially just—as is the case, for example, with overly modest energy access targets for the world's poor.

Improving global resilience to climate variability and hazards will require international efforts to reduce exposure and vulnerability through broad-based adaptation,

---

[11] Roger Pielke Jr., Gwyn Prins, Steve Rayner, and Daniel Sarewitz, "Lifting the taboo on adaptation," *Nature*, Vol. 445 (8 Feb. 2007): pp. 597-598.

shifting focus away from whatever small role anthropogenic climate change presently plays in those hazards.

Adapting to Increased Exposure by Reducing Vulnerability

With global population growth, more accumulated wealth, and other socioeconomic changes, the number of people and amount of property exposed to and thus potentially vulnerable to climate risk will continue to increase, regardless of anthropogenic climate change.[12] Societies can do a better or worse job in maximizing their resilience to climate-related risks. We think a much better job is possible, and history shows this to be the case. Over at least the past century, humanity as a whole has dramatically reduced mortality from natural catastrophes. This success in preserving human life in an often-capricious and frequently harsh environment is the result of innovative adaptations.

Following the IPCC, we define adaptation as adjusting to or preparing for the current or anticipated climate (not just the human-caused change) and its impacts, with the intention of both reducing vulnerability to harm and taking advantage of opportunities. Reducing vulnerability (or, alternatively, improving resilience—not quite the same, but adequate as synonyms for our purposes) is the ability of coupled human and natural systems and their constituent parts "to anticipate, absorb, accommodate, or recover from the effects of a hazardous event in a timely

---

[12] Laurens M. Bouwer, "Have Disaster Losses Increased Due to Anthropogenic Climate Change?" *Bulletin of the American Meteorological Society*, DOI 10.1175/2010BAMS3092.1 (27 Jul. 2010): pp. 39-46.

Figure 5: Annual Global Death Rate (per 100,000) per Decade from Natural Catastrophes, 1900-2013

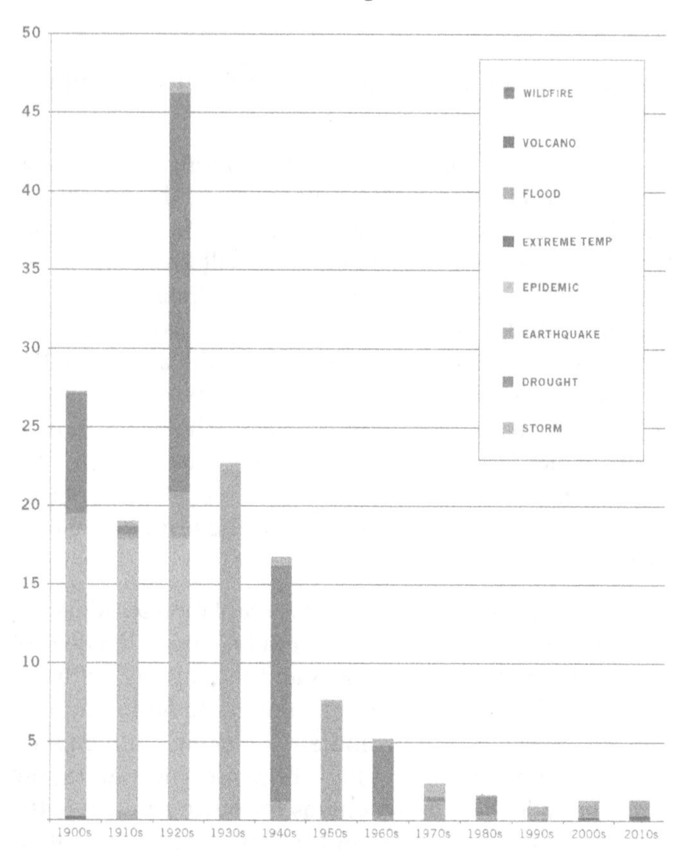

Source: Max Roser, OurWorldInData.org.

and efficient manner."[13] These may be sudden and relatively unpredictable events, like tsunamis or cyclones, or

---

[13] IPCC, "Summary for Policymakers," in *Managing the Risks of Extreme Events and Disasters of Advance Climate Change Adaptation*, C. B. Field, V. Barros, T. F. Stocker, D. Qin, J. J. Dokken, K. L. Ebi, M. D. Mastrandrea, K. J. Mach, G.-K. Plattner, S. K. Allen, M.

slow-onset changes, as with atmospheric warming or changes in rainfall patterns. Effective adaptation efforts intervene in human and natural systems to manage exposure and reduce vulnerability to hazards. Adaptation can be pursued at all levels of governance and scales of action.

Tignor, and P. M Midgely, eds. (New York, NY: Cambridge University Press, 2012), p. 3.

11

EFFECTIVE ADAPTATION STRATEGIES

Adaptation as Institutional Innovation: Crop Development in Nepal

Nepal provides a good example of a novel multilevel institutional partnership that supports adaptation—in this case alleviating the country's food security challenge. The collaboration among farmers, non-governmental organizations, and the government of Nepal works at all stages of technological innovation in agriculture. This includes goal-setting, sharing knowledge with various stakeholders, and developing farmer-preferred technologies suitable to the local environment.[1]

The partnership has been enhanced over the last two decades through an approach called participatory technology development (PTD). While forging dialogue among farmers, researchers, and agricultural policymakers, this collaboration has been vital in setting a research agenda and developing locally appropriate technologies. This novel institutional arrangement has worked to eve-

---

[1] N. Chhetri and W.E. Easterling, "Adapting to climate change: Retrospective analysis of climate technology interaction in rice based farming systems of Nepal," *Annals of the Association of American Geographers* 100, no. 5 (2010): pp. 1-20.

ryone's advantage: it has been cost-effective, location-specific, and met the needs of both farmers and researchers.

This plant breeder works with farmers in Nepal to produce hardier crops for cultivation. *(Photo credit: Netra Chhetri.)*

The outcome of the PTD approach has been the development of hardy rice varieties in a region of Nepal that suffers perennially from poor production due to variable climatic conditions. Institutional and technological innovations of this kind contribute not only to food security but also to climate adaptation. The partnership has enhanced the institutional and technological skills of farmers and communities in plant breeding, seed production, and marketing.[2] This kind of collaboration, facilitated by

---

[2] Bhuwon Sthapit, Abishkar Subedi, Pitambar Shrestha, Pashupati Chaudhary, Pratap Shrestha, and Madhusudan Upadhyay, "Practices supporting community management of farmers' varieties," in *Farmers, Seeds, and Varieties: Supporting Informal Seed Supply in Ethiopia*, M. H. Thijssen, Z. Bishaw, A. Beshir, and W. S.

community-based institutions, allows knowledge to flow between key groups—breeders gain information about the farmers' preferences for specific traits, and farmers learn about and experiment with improved varieties.[3]

Innovative institutional arrangements like the PTD program develop new information and technology aimed at improving resilience; coordinate with social groups and individuals; and can provide financial and leadership support that strengthens local institutional capacities.[4] Today, these arrangements are crucial contributors to Nepal's capacity to recover from the tremendous destruction of the 2015 earthquakes.

Adaptation as Management Strategy: Staying Dry in the Netherlands

The Dutch live with the constant threat of inundation, and have been engaged with an ongoing process of adaptation for more than a thousand years. They have essentially carved their country out of the North Sea through engineering prowess, innovative institutions, and collective action.[5] Future challenges for the country include rising sea levels, sinking land, and changing precipitation

---

de Boef, eds. (Wageningen, Netherlands: Wageningen International, 2008).

[3] N. Chhetri, P. Chaudhary, P. R. Tiwari, and R. B. Yadow, "Institutional and technological innovation: Understanding agricultural adaptation to climate change in Nepal," *Applied Geography*, Vol. 33 (2012): pp. 142-150.

[4] Arun Agrawal, *The Role of Local Institutions in Adaptation to Climate Change*, paper prepared for the Social Dimensions of Climate Change, Social Development Department (Washington, DC: The World Bank, 5-6 Mar. 2008).

[5] Rutger van der Brugge, Jan Rotmans, Derk Loorback, "The transition in Dutch water management," *Regional Environmental Change*, Vol. 5 (2005): pp. 164-176.

patterns.[6] But these risks are well understood, and the country has the adaptive capacity to deal with them for the foreseeable future.

What is more uncertain is Dutch social and economic development, which can change exposure to risk in the long term and which is hard to predict. As in many other affluent countries, the dangers stem not only from the sea or climatic variability, but also from economic changes, demographic shifts, and land-use policies.[7] Recognizing the hazard is not enough: understanding how to address exposure and vulnerability to these hazards is needed for adaptation, and Dutch policymaking and governance has excelled at this.[8]

The Dutch have used their considerable financial resources to reduce the danger posed by flooding, storms, and the ongoing concentration of wealth in vulnerable areas. The Dutch Delta Programme, which addresses these challenges, has its own government commissioner, its own legal basis in the Delta Act, and an annual budget of one billion euros. Much of the protection against flooding involves a capital-intensive system of dikes, dams, seawalls, and other protective infrastructure. But reclaiming and protecting land from the sea also necessitates investment in technological and institutional innovation. Local water boards were organized as long ago as the 13th

---

[6] Paul J. M. van Steen and Piet H. Pellenbarg, "Water Management Challenges in the Netherlands," *Tijdschrift voor Economische en Sociale Geografie*, Vol. 95, No. 5 (2004): pp. 590-598.

[7] Frans Klijn, Karin M. de Bruijn, Joost Knoop, Jaap Kwadijk, "Assessment of the Netherlands' Flood Risk Management Policy Under Global Change," *AMBIO*, Vol. 41 (2012): pp. 180-192.

[8] Martinus J. Vink, Daan Boezeman, Art Dewulf, and Catrien J. A. M. Termeer, "Changing Climate, Changing Frames: Dutch Water Policy Frame Developments in the Context of a Rise and Fall of Attention to Climate Change," *Environmental Science & Policy* 30 (2013): pp. 90-101.

century as a form of democratic governance that allows citizens to make decisions about how best to confront collective risks.[9] These institutions improve not just the capacity to deal with water, but also inspire other government agencies to decide on adaptive practices, such as how to increase the already extraordinary reliability of the Netherlands' power grid.[10]

One of the ways the Dutch build flexibility into their policies and infrastructure is to focus on current or near-term needs but with, when feasible, the option to adapt their protections to future conditions if needed. Thus, for example, levees are constructed or strengthened to withstand existing flood predictions, but are also designed to be heightened in the future, as increasing risks or higher safety standards might necessitate. This strategy is complemented with a systematic lowering of river levels, for instance by means of creating bypasses, or "room for the river." For Dutch water managers and the public, this kind of flexibility is predicated both on the likelihood of a warmer future and the need to deal with the current complexities of a highly unpredictable world.

Adaptation as Disaster Response: Learning from Cyclones in India

A super cyclone slammed into the northeast coast of India in October 1999. Accompanied by tidal surges and days of torrential rain, the eye of Cyclone 05B made landfall in the state of Odisha.[11] Neither officials nor citizens in

---

[9] Paul J. M. van Steen and Piet H. Pellenbarg, "Water Management Challenges in the Netherlands," *Tijdschrift voor Economische en Sociale Geografie*, Vol. 95, No. 5 (2004): pp. 590-598.

[10] Eric Klinenberg, "Adaptation: How can cities be 'climate-proofed'?" *The New Yorker* (7 Jan. 2013): pp. 32-37.

[11] Until 2011, the state of Odisha was known as Orissa. The storm is also referred to as the Paradip Cyclone.

the storm's path took warnings about the impending storm seriously.[12] In the hardest-hit areas, coastal deforestation to make room for prawn fisheries and migrant housing meant there was little to hinder winds with top speeds of 160 miles per hour and tidal surges of 26 feet.[13] The devastation caused by the super cyclone is difficult to comprehend. Official estimates place the death toll at nearly 10,000 people, with millions made homeless; crops and livestock over a wide swath of the state were destroyed, along with the livelihoods they supported.[14] Essential infrastructure like roads and bridges lay in ruins. Relief workers pulled bodies from the mud for weeks afterward.

In 2013, another massive cyclone developed off the east coast of India, in the Bay of Bengal. Cyclone Phailin had worryingly similar characteristics to the 1999 storm: the day before it made landfall, the cyclone had sustained winds of 160 miles per hour, and it smashed into Odisha only slightly further south than the previous storm. Tidal surges destroyed fishing boats and heavy rains caused extensive flooding inland. The high winds swept away houses, uprooted trees, and wrecked power grids throughout the state.[15] But remarkably, of the 13 million people affected by the cyclone, only 44 died: 21 from the

---

[12] Satya Prakash Dash, "How Odisha Managed the Phailin Disaster," *Economic & Political Weekly*, Vol. 48, No. 44 (2 Nov. 2013).

[13] Saudamini Das, "Examining the Storm Protection Services of Mangroves of Orissa during the 1999 Cyclone," *Economic & Political Weekly*, Special Report, Vol. 45, No. 24 (11 Jun. 2011): pp. 60-68.

[14] T. K. Rajalakshmi, "Some lessons from Orissa," *Frontline*, Vol. 18, No. 5 (Mar. 2001).

[15] Satya Prakash Dash, "How Odisha Managed the Phailin Disaster," *Economic & Political Weekly*, Vol. 48, No. 44 (2 Nov. 2013).

cyclone itself (mostly from falling branches), and 23 killed in flash flooding in the storm's aftermath.[16]

A multipurpose shelter like this one protected citizens of Odisha from the worst effects of Cyclone Phailin in 2013. *(Photo credit: ADRA India.)*

Why did vastly fewer people die during Cyclone Phailin than Cyclone 05B—or, for that matter, from other notable storms like Typhoon Haiyan (with more than 6,300 deaths), Hurricane Katrina (more than 1,700 people), and Hurricane Sandy (285 fatalities)? The most important answer is that in the years after the devastation of 1999, the state of Odisha and its coastal communities resolved that storm casualties were unacceptable. This provided a powerful motivation for effective protection measures, and was the inspiration for the adaptation goal proposed in this report.

[16] Lindsey Harriman, "Cyclone Phailin in India: Early warning and timely actions saved lives," UNEP Global Environmental Alert Service (Nov. 2013).

In cooperation with the Indian central government and donors like the World Bank, these communities put tremendous effort into reducing their vulnerability to another super cyclone. Adaptation initiatives included building cyclone shelters as multi-purpose structures that house schools when not being put to emergency use (ensuring their upkeep); developing contingency plans for evacuating and housing hundreds of thousands of coastal residents; and improving the Indian Meteorological Department's storm tracking and predictions to enable accurate early warnings.[17] Similar examples of responsive governance leading to fewer deaths from storms can be found in other parts of South Asia.[18]

Adaptation as Sharing Knowledge: Preparing for Tsunamis

Effective adaptation, like other innovation processes, requires learning from past experience, experimenting with novel practices and technologies, and participating in a network of people who can share and evaluate new practices that emerge from particular contexts. During the 2004 Indian Ocean tsunami, for example, which killed nearly 230,000 people, inhabitants of Simeulue Island in Indonesia knew to head into the mountains when they saw the ocean retreating, through knowledge passed down from the generation that had experienced a deadly tsunami in 1907. They survived in much greater numbers

---

[17] Saurabh Dani, "Never Again! The Story of Cyclone Phailin," End Poverty in South Asia, World Bank Blog (21 Oct. 2013).

[18] P. Peduzzi, U. Deichmann, A. Maskrey, F.A. Nadim, H. Dao, B. Chatenoux, C. Herold, A. Debono, G. Giuliani, and S. Kluser, "Global Disaster Risk: Patterns, Trends and Drivers," in *Global Assessment Report on Disaster Risk Reduction* (Geneva, Switzerland: United Nations International Strategy for Disaster Reduction, 2009), pp. 17-57.

than other places in Indonesia, whose inhabitants may not have had that knowledge. Several programs since then have sought to educate coastal communities about what to do for the next tsunami, using social networks to provide information, rather than high-tech measures that fail to reach everyone and have been unreliable.[19]

A conceptual drawing for an elevated Tsunami Evacuation Park (TEP) in Padanh, Indonesia. *(Image credit: Kornberg Associates.)*

It may be impossible for coastal communities, particularly in densely populated cities, to reach higher ground even if adequately warned of an approaching tsunami. In a neat reversal of traditional evacuation plans, one innovative approach seeks to bring higher ground to low-lying communities in the form of elevated parks. A half-dozen elevated parks in the Indonesian city of Padang could save as many as 100,000 people from the threat of inundation. Simple, relatively inexpensive, and reassuring to the city's inhabitants, this innovation depends more on knowledge of a particular social and institutional context

---

[19] Julie Morin, "Tsunami-resilient communities' development in Indonesia through educative actions," *Disaster Prevention and Management*, Vol. 17, No. 3 (2008): pp. 430-446.

than of the science of tsunamis or complex models of how they work.[20]

Lesson: To Adapt for Rising Exposure to Climate Change, Innovate toward a Range of Possible Futures

The one sure thing about the future is that the unexpected will occur. Dealing with uncertainty requires flexibility and foresight to keep pathways open.

An adaptation agenda, with the ambitious goal of minimizing harm from natural disasters, means setting goals with concrete benefits now—but keeping an eye toward a variety of potential futures. The pathways for achieving the goals that communities want and need for adaptation are varied. For example, increasing risks of flooding due to sea level rise or higher peak river discharges can be approached with elaborate infrastructure, by retreating from vulnerable areas, by improving evacuation procedures, by flood-proofing buildings and infrastructure, and through democratic participation in a process that decides how the country will develop socioeconomically.

Even as nations develop economic and infrastructural resilience to extreme events, there are better and worse approaches to adaptation. The above case studies demonstrate locally relevant, forward-looking, flexible policies to increase resilience even where significant capital and human life are exposed.

---

[20] Daniel Sarewitz, "Brick by brick," *Nature*, Vol. 465 (6 May 2010): p. 29.

12

VULNERABILITY REDUCTION

Vulnerability to climate extremes is not equitably distributed. Reduced vulnerability is highly correlated with both individual and societal wealth. Wealthy societies can afford infrastructure and institutions that poor societies often cannot, from sea walls, flood channels, and evacuation shelters to emergency response systems, building codes, and the capacity for implementation. With higher incomes come air conditioning, refrigeration, and communication technologies that help individuals navigate both immediate crises and chronic long-term shifts in the natural environment.[1] Prosperous societies are able to invest in institutions and decision-making processes that embody principles of good governance, providing a solid foundation for identifying adaptation priorities, making fair trade-offs, and building resilience.[2]

Therefore the most effective way to expand options for dealing with both expected and—often more importantly—unexpected changes is through socioeconomic development. In general, economically distressed and

---

[1] Subhro Niyogi, "Row over activist says AC a must in Kolkata," *Times of India* (24 Apr. 2015).

[2] M. Wagner, N. Chhetri, M. Sturm, "Adaptive capacity in light of Hurricane Sandy: The need for policy engagement," *Applied Geography*, Vol. 50 (2014): pp. 15-23.

disenfranchised communities are fragile and vulnerable to hazards, with few resources to invest in protection or recovery from extreme events. In terms of governance and institutions, the weaker and less democratic they are, the more citizens suffer when natural disasters occur. Economic development and responsive governance provide a strong foundation for dealing with all kinds of unpredictability, from epidemics to storms to economic recessions.[3] Poor populations are generally not more vulnerable to climate risk than rich countries because of their geographies, but because they have not yet acquired the resilience advanced by socioeconomic development processes.

Adaptation as Driver of Development: Energy Access in Sub-Saharan Africa

The conspicuous lack of access to modern energy sources in sub-Saharan countries has resulted in a staggering 80 percent of urban households using charcoal for cooking and heating. Charcoal production to meet this need has created swathes of deforested land around cities like Addis Ababa, Lusaka, and Kampala.[4] Deforestation has a number of negative consequences, ranging from soil erosion (which exacerbates flooding and landslides), to poor agriculture (and food insecurity), to increased risk of wildfire (because secondary growth forests are more susceptible to fire). It also destroys an important source of atmospheric carbon sequestration. In addition to these

---

[3] Matthew E. Kahn, "The Death Toll from Natural Disasters: The Role of Income, Geography, and Institutions," *The Review of Economics and Statistics* Vol. 87, No. 2 (2005): pp. 271-284.

[4] Werner L. Kutsch et al., "The Charcoal Trap: Miombo forests and the energy needs of people," *Carbon Balance and Management*, Vol. 6 (2011); and Leo C. Zulu and Robert B. Richardson, "Charcoal, livelihoods, and poverty reduction: Evidence from sub-Saharan Africa," *Energy for Sustainable Development* (2012).

regional impacts, charcoal's use indoors causes considerable health problems at the household level.

The Katse Dam in Lesotho is an important source of water and energy. *(Photo credit: Christian Wörtz.)*

Access to reliable modern energy services can reduce vulnerability to these hazards: powered irrigation systems make farmers less reliant on unpredictable rainfall; effective healthcare depends on energy services like electricity and fuel for refrigeration, transportation, and device operation; and abundant energy drives industrial expansion, urbanization, and job growth. Equitable access to reliable and abundant energy also reverses the enormous environmental despoliation created by the subsistence living of hundreds of millions of poor people. This despoliation itself greatly magnifies the risk of natural disasters, such as mudslides and floods caused by deforestation of mountainous regions.

Without reliable, modern forms of energy like electricity and natural gas, many communities in sub-Saharan Africa are deprived of the opportunities for socioeconomic advancement presented by energy access. Forced to meet cooking and heating needs with fuels like charcoal, energy poverty exacerbates inequality—particularly for the women and children who spend a considerable amount of their time finding fuel and cooking indoors—and vulnerability to many hazards. The minimal or nonexistent energy access in developing countries becomes an example of maladaptation not just to future challenges, but current conditions as well.

Yet, as we explored previously, the kind of energy sector development that underpins prosperous and resilient societies lies outside, and in many ways directly contradicts, the framework within which energy, development, and climate change are typically linked. Reorienting that framework toward the imperative of improving access to abundant, affordable, and increasingly clean energy provides a much sturdier foundation for development, resiliency, and environmental protection. This should be the "win-win" priority for organizations and governments that seek to foster sustainable development.

Adaptation as Greater Equality: Public Health and Extreme Heat

The individuals most vulnerable to a range of hazards tend to be those with the weakest social networks within their communities—the old, very young, infirm, poor, or otherwise marginalized.[5] Advance warning of dangers do little to help these individuals, and they often lack the ability to access even basic services (like transportation, healthcare, public shelters, or even information about natural disasters) that can help them cope with extreme events.

---

[5] Kenneth Hewitt, *Regions of Risk: A Geographical Introduction to Disasters* (New York, NY: Routledge, 1997).

Improved social bonds can help communities deal with hazards like heatwaves. *Photo credit: Vasilios Sfinarolakis.)*

The deadly 1995 heat wave in Chicago, Illinois provided ample evidence of this kind of vulnerability. A disproportionate number of the 739 people who died from the heat were elderly and African American, living alone in poor, violent, segregated neighborhoods. But not all communities with these characteristics suffered equally. In demographically similar neighborhoods, but ones where businesses had not fled and community organizations remained active, people survived the heat wave at rates comparable to or exceeding more affluent areas of the city. Churches, block clubs, and especially commercial activity provided an informal social support network that ensured many fewer people died during the city's worst natural disaster.[6] A similar story played out in Europe during its deadly heat wave in 2003.

Poor and marginalized individuals may remain isolated and alone during other crises, such as flooding, hurri-

---

[6] Eric Klinenberg, *Heat Wave: A Social Autopsy of Disaster in Chicago* (Chicago, IL: University of Chicago Press, 2003).

canes, earthquakes, or influenza outbreaks.[7] Improving social bonds among these individuals and with the broader community thus offers a powerful form of resiliency that improves the overall health of individuals and communities and their development prospects. The benefits of strong social capital, underpinned in part by a vibrant commercial sector, range from reduced obesity and diabetes rates,[8] to lower levels of crime,[9] to longer life spans.[10] With these impacts in mind, over the past fifteen years the Centers for Disease Control and Prevention has launched community engagement programs in various American cities to build and strengthen the kinds of local social infrastructure that help vulnerable communities deal with hazards and address public health concerns like smoking, cancer, and heart disease.[11]

---

[7] Christopher R. Browning, Seth L. Feinberg, Danielle Wallace, and Kathleen A. Cagney, "Neighborhood Social Processes, Physical Conditions, and Disaster-Related Mortality: The Case of the 1995 Chicago Heat Wave," *American Sociological Review*, Vol. 71, No. 4 (2006): pp. 661-678.

[8] CTSA Community Engagement Key Function Committee Task Force, *Principles of Community Engagement*, 2nd Ed. (Washington, DC: National Institutes of Health, 2011).

[9] Robert D. Putnam, *Bowling Alone: The Collapse and Revival of American Community* (New York, NY: Simon & Schuster, 2001).

[10] Eric Klinenberg, "Adaptation: How can cities be 'climate-proofed'?" *The New Yorker* (7 Jan. 2013): pp. 32-37.

[11] CTSA Community Engagement Key Function Committee Task Force, *Principles of Community Engagement*, 2nd Ed. (Washington, DC: National Institutes of Health, 2011).

Lesson: High-Energy Adaptation Means Reduced Vulnerability

Reducing vulnerability to natural disasters depends on prioritizing social and economic development. Adaptive capacity requires more energy for climate control in buildings, high-tech crops and energy-intensive fertilizers, levies and sea walls, and the kinds of commercial activities that improve social bonds. Successful adaptation will occur only on a high-energy planet.

While the correlation between socioeconomic development and climate resilience is strong, it is important to recognize that causation runs both ways. Improved adaptation to climate extremes and natural disasters brings greater socioeconomic development, even as that development improves resilience. Air conditioning in tropical countries, for instance, not only makes the population less vulnerable to heat waves, but also dramatically increases labor productivity year round.[12]

How best to accelerate socioeconomic development remains a vexed and controversial topic. We simply argue that fostering development prospects and decreasing disaster vulnerability are synergistic. Ambitious adaptation objectives—as India's declaration to eliminate deaths from cyclones demonstrates—can be powerful drivers of these development processes, even in countries that are not yet affluent.

Prioritizing human development as an adaptation strategy requires actions and investments that are most appropriate within that development context, rather than the conventional mitigation focus of climate policy. For example, achieving more access to modern energy for more people, which is a problem from the mitigation perspective, is a solution pathway from an adaptation per-

---

[12] Subhro Niyogi, "Row over activist says AC a must in Kolkata," *Times of India* (24 Apr. 2015).

spective. A good first step to improving adaptation in developing countries would therefore be to address the vast inequities in access to modern energy. In previous chapters we explained how a focus on energy access and modernizing energy production and distribution systems can accelerate technological innovation that leads to cleaner, cheaper energy. Here we emphasize that energy access also helps build resilience into all societies—an approach in which mitigation and adaptation are complementary.

13

REIMAGINING CLIMATE ACTION

The effectiveness of an adaptation agenda should be measured in lives protected and livelihoods saved. The priority for such an agenda is the progressive and continual reduction of average number of deaths each year from natural disasters, including those disasters that will be exacerbated by a changing climate. With actions that are attentive to peoples' well-being, equity, and livelihoods, adaptation policies focus on opportunity rather than cost, promising benefits that are near-term and certain rather than long-term and uncertain. And as the examples discussed here illustrate, reduced vulnerability to natural disasters is something societies have succeeded at achieving.

Adaptation activities neither preclude nor discourage—and in fact should complement—action on the critical task of reducing GHG emissions. Socioeconomic development and energy innovation open new opportunities for stabilizing atmospheric greenhouse gas concentrations. Energy innovation is also a broadly adaptive process because the prosperous modern societies enabled by improved energy access and innovation capacity are better able to deal with hazards, for reasons ranging from better infrastructure to more effective public institutions.[1] Localized adaptations also support mitigation efforts. One example is urban planning that prioritizes high-density, mixed-income neighborhoods, mass transit, and infrastructure produces urban communities that are both climate resilient and lower-emission on a per capita basis.[2]

Social institutions are a critical component of climate adaptation.[3] Nepal's farmer-managed irrigation systems, for example, perform a range of activities, including pooling resources for maintaining irrigation waterways, regulating water distribution and allocation, monitoring rule violation, and arbitrating and negotiating conflict. These local institutions have historically been instrumental in safeguarding resources, including the protection of forests and watersheds. In recent decades, these institutions have played an important role in bridging the gap between researchers, development workers, and the lay community. Such local institutions are vital in facilitating climate adaptation processes, whether in the foothills of the Himalayas or inner-city Chicago. Yet, as we saw in the aftermath of Nepal's earthquakes in 2015, communities are threatened by—and thus must be resilient to—a range of hazards, not just those related to climate.

These and other adaptations are not necessarily cutting-edge, nor are they directed only at the anticipated impacts of a warming climate, and that is our point. What

---

[1] Matthew E. Kahn, "The Death Toll from Natural Disasters: The Role of Income, Geography, and Institutions," *The Review of Economics and Statistics*, Vol. 87, No. 2 (2005): pp. 271-284.

[2] Vishaan Chakrabarti, "How Density Makes Us Safer During Natural Disasters," *The Atlantic Cities* (19 Sep. 2013).

[3] N. Chhetri, P. Chaudhary, P. R. Tiwari, and R. B. Yadow, "Institutional and technological innovation: Understanding agricultural adaptation to climate change in Nepal," *Applied Geography*, Vol. 33 (2012): pp. 142-150.

these adaptations have in common is that they build on a long history of adjusting to dynamic environments and improving well-being in the context of current and expected challenges. They do not depend, as action under the UNFCCC does, on determining which phenomena are the result of "dangerous anthropogenic interference with the climate system." They do not even depend on prioritizing climate change as the core issue, although this may be an effective entry point for addressing problems such as sea level rise.

Crises—and the increased possibility of crises—offer opportunities for increasing resilience. Because of their immediacy, they can unite communities around the need for resilience in ways that abstract or future-oriented climate predictions cannot. In addition, the impact of extreme events offers metrics for evaluating adaptation efforts: Did a community's adaptations produce resiliency when tested by events? Conversely, reduced vulnerability provides longer-term benefits beyond disasters themselves. The initiatives described in Odisha and the Indian Ocean communities have the potential to create positive feedbacks of effective governance, democratic participation, and increased adaptation, as the public demands an end to avoidable deaths and the government implements programs to realize that goal.

The options offered by an adaptation agenda are also necessary from a political perspective: unlike mitigation's gridlock and divisiveness, adaptation empowers decision makers to act on a variety of important issues in ways that resonate with their constituencies. In some contexts, that will mean explicitly connecting policies to addressing climate change. In other cases, action can be premised on concerns like raising living standards for those in poverty or protecting lives and property against natural hazards of all kinds. The politics of adaptation align with the values of virtually all citizens.

Pragmatic Action

Climate policy writ large, and international climate negotiations in particular, have prioritized emissions reduction and put adaptation on the back burner. In our view, hanging the long-term "solutions" for everything from food security to public health to economic growth on how well the world can reduce its greenhouse emissions is untenable and ineffective. Such an approach makes it difficult to see the advantages of dealing pragmatically with impacts of climate in the present, through adaptive strategies that can deliver near- and medium-term benefits that justify the political and economic costs of taking action.

Politically, the mitigation approach has framed climate change as a narrow problem with a linear relation between action and impact. *Reduce greenhouse gas emissions to avert catastrophe!* is a simple formula that translates nicely into bumper stickers such as "Save the Planet" or "Stop Global Warming." It defines all action that connects to reduced emissions as desirable, and all others as bad. It specifies goals mainly in terms like parts per million of atmospheric carbon dioxide and its benefits reside in an unknowable future. In the real world, GHG concentrations are but one abstract manifestation of the complex sociotechnical systems on which humans depend for their well-being. And despite the enormous efforts expended to create a global greenhouse gas mitigation regime, the result has mostly been political divisiveness—and, so far, ever-rising emissions.

In stark contrast, adaptation to the current and anticipated climate (and not just its human-caused changes) is a process of identifying multiple pathways for achieving concrete goals like reduced deaths and economic losses from disasters, more equitably distributed prosperity, stronger communities, better public health, and more access to modern energy systems.

The National Center for Research on Earthquake Engineering in Taipei, Taiwan, can perform full-scale testing of earthquake-resistant building designs. *(Photo credit: Gatutigern.)*

An adaptation agenda includes an array of activities beneath its broad umbrella, but central to it is creating options and opportunities for achieving desirable outcomes under a range of possible futures.[4] Determining which outcomes are desirable and what futures are possible is the messy task of our democratic institutions; achieving this adaptability through continual learning is the responsibility of our communities and organizations. Positive, pragmatic adaptive action—difficult and incremental as it may sometimes be—is a powerful force for protecting human dignity, livelihoods, and prospects in the face of a dynamic climate and rapidly evolving societies. It is pluralistic and inclusive, it promises near-term benefits for near-term costs, and it offers broad opportuni-

[4] Roger A. Pielke Jr., Daniel Sarewitz, and Radford Byerly Jr., "Decisionmaking and the Future of Nature: Understanding and Using Predictions," Ch. 18 in *Prediction: Science, Decision Making, and the Future of Nature*, Daniel Sarewitz, Roger A. Pielke Jr., and Radford Byerly Jr., eds. (Washington, DC: Island Press, 2000).

ties for political support. Nearly everyone can see themselves as a potential "winner" in some aspect of an effective adaptation agenda.

Innovation, a process through which new (or improved) technologies and institutions are developed and brought into widespread use, has always been an integral part of the social response to the multiple stresses from the natural world. Far from being a new activity undertaken in response to anthropogenic climate change, humans excel at the kinds of innovation-led adaptation which have lessened vulnerability to hazards, and allowed humans to flourish in an incredible diversity of climates.[5] Technological examples include food preservation techniques to overcome the problem of seasonal shortages; aluminum and other structural materials that resist environmental deterioration; antifreeze to safeguard internal combustion engines in the winter; and weather and earth resource satellites for analysis of weather and climate. Other adaptations include vaccine development, property insurance, disaster preparedness, forest management, and biodiversity conservation—the list is nearly endless.

To unleash this force we emphasize in particular two focusing strategies. The first is to adopt progressive and decisive reductions in loss of life from disasters worldwide as a direct measure of adaptive success. The empowering lessons of two regions as socioeconomically distinct as Odisha and the Netherlands show that such a goal can be within reach for all nations and people. The second strategy is to put improved adaptation to climate at the center of the international policy agenda, along with energy access and innovation. This new, pragmatic focus can create a multitude of political opportunities and poli-

---

[5] Jesse H. Ausubel, "Does Climate Still Matter?" *Nature*, Vol. 350 (25 Apr. 1991): pp. 649-652.

cy pathways for emerging from a quarter century of climate change gridlock into an era of enhanced human thriving and improved capacity to manage the natural systems that sustain us.

LIST OF ACRONYMS

AGECC	United Nations Advisory Group on Energy and Climate Change
AR5	Intergovernmental Panel on Climate Change Fifth Assessment Report
BRICS	Brazil, Russia, India, China, and South Africa
CATF	Clean Air Task Force
CCS	Carbon capture and storage (or sequestration)
CDM	Clean Development Mechanism
CIF	Climate Investment Funds
DOE	United States Department of Energy
EOR	Enhanced oil recovery
GEF	Global Environment Facility
GHG	Greenhouse gas
IEA	International Energy Agency
INDC	Intended Nationally Determined Contributions
IPCC	Intergovernmental Panel on Climate Change
ITC	Investment tax credit
kWh	Kilowatt hour
MSR	Molten salt reactor

MW	Megawatt
OECD	Organisation for Economic Co-operation and Development
OPIC	Overseas Private Investment Corporation
PTD	Participatory technology development
PV	Photovoltaic
R&D	Research and development
SE4All	Sustainable Energy for All
SIBRATEC	Sistema Brasileiro de Tecnologia
SREX	Special Report on Managing the Risks of Extreme Events and Disasters to Advance Climate Change Adaptation
UN	United Nations
UNFCCC	United Nations Framework Convention on Climate Change
USAID	United States Agency for International Development

ACKNOWLEDGEMENTS

We wish to express our gratitude to Irfan Ali, John Alic, Armond Cohen, Sharlissa Moore, Gregory Nemet, Megan Nicholson, Aneri Patel, Tobias Schmidt, David Spielman, Matthew Stepp, Christine Sturm, Letha Tawney, Paul Wilson, T. Stephen Wittrig, and all the participants at the three Climate Pragmatism workshops for their input, feedback, and critiques of the reports on which this book is based. Breakthrough Generation fellows Erik Funkhouser and Oliver Kerr contributed foundational research and analytics to the *High-Energy Innovation* report. We also thank Jenna Mukuno of the Breakthrough Institute for editorial assistance, Felicity O'Meara for proofreading, Erin Aigner for map design, Dita Borofsky for graphic design and layout, and series editor G. Pascal Zachary for guidance on turning this into a book.

Finally, we express our sincere appreciation for the generous support of Peter Teague and the Nathan Cummings Foundation, without which the Climate Pragmatism project would not have been possible.

CPSIA information can be obtained
at www.ICGtesting.com
Printed in the USA
BVHW042144160420
577787BV00009B/189